THE PORCELAIN GOD

THE PORCELAIN GOD

A SOCIAL HISTORY
OF THE TOILET

JULIE L. HORAN

Illustrations by Deborah Frazier

A CITADEL PRESS BOOK
Published by Carol Publishing Group

For John Muraca
and Mary Kaye Collins

Thank you

Carol Publishing Group Edition, 1997

Illustrations by Deborah Frazier

A Citadel Press Book
Published by Carol Publishing Group
Citadel Press is a registered trademark of Carol Communications, Inc.

Editorial, sales and distribution, and rights and permissions inquiries
should be addressed to Carol Publishing Group, 120 Enterprise Avenue,
Secaucus, N.J. 07094.

In Canada: Canadian Manda Group, One Atlantic Avenue, Suite 105, Toronto,
Ontario M6K 3E7

Carol Publishing Books are available at special discounts for bulk
purchases, sales promotion, fund-raising, or educational purposes. Special
editions can be created to specifications. For details, contact: Special Sales
Department, Carol Publishing Group, 120 Enterprise Avenue, Secaucus, N.J.
07094

Manufactured in the United States of America
10 9 8 7 6 5 4 3 2 1

The Library of Congress has cataloged the Birch Lane Press edition as
follows:

Horan, Julie L.
     The porcelain God : a social history of the toilet / Julie L.
Horan.
          p.   cm.
     "A Birch Lane Press book."
     ISBN 0-8065-1947-9 (paperback)
     1. Toilets—History.   2. Toilet paraphernalia—History.
I. Title.
GT476.H67   1996
394—dc20   95-50110
                                                     CIP

CONTENTS

BEFORE I START . . .

The toilet represents the seat of life. Life began in 1500 on the toilet for Charles V, the future Holy Roman Emperor of vast European domains, whose mother gave birth to him while on the privy. And Elvis, who died on one, illustrates that the toilet has been the seat of death [for some]. Although the toilet and its predecessors have served the needs of humanity for hundreds of years, few people have any knowledge of its history. Who invented the first flush toilet? No, it was not Thomas Crapper. What did people use before the invention of the toilet? Why are toilets referred to as the throne?

In the following pages I hope to answer these and other questions. But before I start, I want to explain my sources in case some of you find the facts a little offensive to the various cultures mentioned. Through advertisements, diaries, and scatological works, I have compiled a collection of bizarre stories. Many of my sources come from Western travelers during the age of exploration and the years of the British Empire as they investigated new lands and reported on the practices of the native people they encountered. The accuracy of the accounts rely on the personal interpretations of the various authors. Some of them appear implausible. However, in a few cases, they are supported by other witnesses. I do not wish to offend anyone, only to offer curious habits along with a history of the toilet.

Now for a peek at what's inside. Initially, relief was found in the great outdoors. As humans forged ahead intellectually and socially, sanitation evolved sporadically. Some civilizations possessed toilets more advanced than other developments. Still other societies have shown man as little more than an animal in maintaining cleanliness. In the past, the modus operandi of disposing human waste basically consisted of collecting the offense in a pot and throwing the contents out the window. The predictable repercussions of accumulating filth provide an insightful and comical view of human development. The recent history of sanitation practices, and the development of the toilet, arose from collective national responses to epidemic diseases. The cholera outbreaks in Europe during the mid-nineteenth century, particularly in Britain, greatly advanced improvements in the toilet and its public acceptance.

Conventional historians have concentrated on the deeds of notable men and women. Reviewing toilets throughout history has an equalizing effect on social hierarchies. After all, everyone must pull down his pants or lift her skirt to relieve themselves. Only the facilities differ. King Henry VIII owned a close-stool made of black velvet and studded with two thousand gold nails. The Sun King, Louis XIV, believed it improper to cut short a conversation to "go to the bathroom." To the horror of visitors, Louis would receive them while sitting on the pot. From the third millennium B.C. when the earliest "toilets" were found in the Indus valley to the high-tech twentieth-century toilets of the Japanese, the study of toilets offers an entertaining view of life.

THE
PORCELAIN
GOD

⊰ 1 ⊱

The Dawn of Civilization

Toilet-philes argue that civilization began not with the advent of written language but with the first toilet. Waste control allowed individuals to quit wandering the earth trying to escape their dung and finally settle down. The ancient homelands in Mesopotamia and the Indus valley civilization possessed the first cesspits and sewer systems marking the beginning of what we now call the civilized world. As early as 3300 B.C., Habuba Kabir, an ancient Mesopotamian city in what is now Syria, used pipes to carry off waste water. People of the Indus valley in India used waste water to "flush" privies that emptied into brick-lined cesspits. Egyptians, Greeks, and the masters of masonry—the Romans—all developed sophisticated sanitation systems for their era.

Mesopotamia

Mesopotamia has been called the "cradle of civilization" because complex social institutions first appeared there. But Mesopotamia deserves another title as well: the "seat of sanitation." The inhabitants of Mesopotamia were among the first to address the problems associated with human waste.

The Sumerians of Mesopotamia ruled the region between the Tigris and Euphrates rivers during the third millennium B.C. Their ruler, Sargon I, known as the King of Kings, enjoyed conquering other lands. Sargon's most important achievement was setting an example of cleanliness by building six privies in his palace. An improvement over the widely used earthen pots that required the user to squat precariously, the king's privies provided a sitting space over a cesspit. The early toilet seat resembled a large horseshoe, comfortably fitting around the person's posterior. Despite the inventions of writing and the seated privy, centuries passed before the bathroom experience was completed with the invention of the newspaper.

Indus Valley

The Indus civilization had a history similar to Mesopotamia's. Settling the area that is now Pakistan in the third millennium B.C., the Harappans developed a remarkably advanced society. Excavations of a Harappan settlement in Mohenjo-Daro revealed a modernlike city of bricks strong enough to use in two-story buildings. Many homes were hooked up to a street drain submerged in the ground. Privies and baths emptied into drains that flowed to a number of connecting cesspits. The Harappans covered the street sewers in an attempt to block the strong odor coming from the waste. The Harappans' exceptional city management is placed in perspective by noting that many modern cities around the world continue to rely on *open* sewers.

The Harappans displayed their reverence for cleanliness by building a huge bathtub in the city center. In addition, each private house had a bathing platform that served as a type of shower. A privy was connected to the outside wall of the bathing area, making for the first modern bathroom. The privy, equipped with a brick or wood seat, could provide comfort for long "deliberations." The contents of the privy traveled down a chute into the street sewers. Because some of the chutes did not fully extend to the sewer, the odor must have been offensive. If a house was built back from the road, the chute emptied into a barrel or cesspit, which was then emptied into the sewer.

Minoans—The Island of Crete

Wealth, luxury, and style distinguished the early inhabitants of Crete. Intricate gold masks and elaborate fashions for women indicate a society concerned with aesthetics in daily life as well as the imagined rewards of an afterlife.

The Palace of Knossos, built by the great King Minos, typified the sumptuous atmosphere of the island. Knossos was a massive palace with over 1,400 rooms. Paintings, columned staircases, courtyards, and an intricate labyrinth adorned the palace, but it also contained a feature far more important than elaborate decor: flush toilets. The palace water-supply system consisted of a series of cone-shaped terracotta pipes ingeniously interconnected. A pan on the roof collected rainwater to supply the pipes. Because of the unique shape of the pipes the velocity of rushing water was slowed, preventing overflow. Piped water served the bathrooms and latrines of the palace. The piped water flushed the contents of the latrines in a manner similar to modern flush toilets. A wooden seat covered the opening. An amazing device for its time, the Knossos toilet ensured the aesthetic integrity of the Cretans and their olfactory senses.

Troy

The ancient city of Troy, home to the beautiful Helen, was noted by contemporaries, not for its Greek invaders' Trojan horse, but for its hygiene habits. Captain John G. Bourke of the U.S. Cavalry in the late nineteenth century operated as an amateur anthropologist during his travels among Native Americans. He wrote a fascinating book on the historical role excrement played in Western and non-Western cultures. In *Scatologic Rites of All Nations*, Captain Bourke, commenting on the habits of the ancients, reported that in 1200 B.C. the citizens of Troy defecated in the "full light of day."

The British Museum's antiquities collection reflects strong interest in and knowledge of life in ancient Egypt, Greece, and Rome. Examining the sanitation systems and habits of each of these civilizations expands our knowledge of the "beginning of Western civilization."

Egypt

The Egyptians did little to advance the cause of sanitation. Their greatest contribution was found in the city of Akhenaton at Tel-el-Amarna. The house of a high official, dating back to the fourteenth century B.C., contained a privy built into the recess of the wall behind a bathroom. Made of limestone, the seat of the privy was shaped like a keyhole, similar to those of earlier Mesopotamia. The split seat created a more comfortable fit for the buttocks and anticipated the shape of modern toilet seats. The contents of the privy fell into a removable vase in a pit below.

The writings of the ancient Greek historian, Herodotus (484–425 B.C.), provide valuable information, not only on the political events occurring in Greece but also on the lifestyles of contemporary cultures. During his lifetime Herodotus traveled through Asia Minor and Egypt. While in Egypt, Herodotus characterized Egyptian daily habits as mysterious and confusing. From a selection of his *Histories*, Herodotus recounts his visit to Egypt:

The Egyptians established for themselves manners and customs in a way opposite to other men in almost all matters: for among them the women frequent the market and carry on trade, while the men remain at home and weave; . . . the men carry their burdens upon their heads and the women upon their shoulders; the women make water standing up and the men crouching down; they ease themselves in their houses and they eat without in the streets, alleging as reason for this that it is right to do secretly the things that are unseemly though necessary, but those that are not unseemly, in public.

The pharaohs of ancient Egypt believed they descended directly from the gods. As the link between mortal man and the heavens, the pharaoh was obligated to assist his subjects in gaining entrance to the afterlife by ruling successfully. Drought, disease, and natural disasters tested the deity status of the pharaoh and his ability to remain in power. Convincing others of his superiority required the pharaoh to have strong leadership traits and lots of luck. According to author Reginald Reynolds, one of the pharaohs (it is unknown which one) thought that natural bodily functions too closely linked him with mere mortals. To maintain his status as a deity, the pharaoh would sneak out of his palace before dawn to defecate in the desert while no one was looking.

In ancient Egypt, the disposal of human and animal excrement, carcasses, offal, and other rubbish was handled in a variety of ways. Nature provided the means for getting rid of small amounts of debris. The hot desert sun quickly caused anything left under its rays to disintegrate. Many people would carry their dead animals, collected waste, and even unwanted infants to the edge of the desert for dumping. Others merely dumped the waste into the river. The presence of natural scavengers offered another choice as birds, dogs, and cats wandered the unpaved streets devouring everything in sight. Artwork depicts the presence of these scavengers, but special attention was given to one in particular—the scarab.

Egyptians revered the scarab. The beetle is found on pottery, paintings, and jewelry of the period. The image of the scarab was believed to act as a talisman averting evil and bringing good fortune to the owner. Interestingly, in nature the scarab performs a disgusting, but important, job. The beetle searches for dung and then rolls it around until the dung is spread over a wide area. Acting as the farmer's helper, the beetle assists in fertilizing the land. In his book *Les Lieux*, Roger-Henri Guerrand suggests that the scarab looks remarkably like a clump of dung itself. Could it be that the ancient Egyptians were worshiping excrement? Certainly the scarab and dung provided valuable services to the Egyptians.

During his travels in Egypt, Herodotus reported that an Egyptian king, Amasis, had owned a gold footpan used for washing his feet as well as for collecting vomit and urine. To convince his subjects that he was a god, Amasis secretly turned the footpan into a religious relic. When his subjects worshiped the footpan as an idol, Amasis revealed that the footpan belonged to him. Based on the logic that worshiping the footpan equaled worshiping himself, Amasis declared himself a deity.

Hebrews

The ancient tribe of Moabites, distantly related to the Hebrews, differed remarkably from their cousins. The Moabites worshiped several gods and created bizarre rituals and practices. One of the gods the Moabites worshiped was named Bel-Phegor, the god of dung—a deity often found in agricultural communities. Worshipers wishing to make an offering to Bel-Phegor would drop their pants before the altar and defecate.

The Moabite city of Shittim (an appropriate name) may have been the location of a beastly murder. King of the Moabs, Eglon, was an immensely fat man. One day, while sitting on the privy, he was having a chat with Ehud, one of his subjects. Suddenly, Ehud leapt to his feet and drove a knife into the belly of the king. The knife, completely imbedded, could not be retrieved. Eglon died from his wounds.

Setting the ground rules for cleanliness, God sent the following instructions to his followers, the Hebrews, in Deuteronomy 23:12. "You shall have a designated area outside the camp to which you shall go. With your utensils you shall have a trowel; when you relieve yourself outside, you shall dig a hole with it and then cover up your excrement. Because the Lord your God travels along with your camp, to save you and to hand over your enemies to you, therefore your camp must be holy, so that he may not see anything indecent among you and turn away from you."

Adhering to God's desire for a clean environment, many Jewish households contained primitive privies. A stone privy seat with a six-inch hole was found in Jerusalem dating from the end of the Iron Age. Sanitation was valued and so was the privy. (According to author Reginald Reynolds, the Jews called the privy "the house of honor," at a later time in their history, starting in the sixteenth century.) Jerusalem included a gate on the southeast side of the city marking the area where waste was to be dumped, to be carried away by the river Cedron. Fittingly, it was named Dung Gate. Lazy citizens began using an alternative area of the city as a community dunghill, unmoved by the fact they were committing blasphemy. The dunghill grew to an unbearable size. Holy men believed the dunghill represented the torments one could expect to find in hell. Early Christians mistakenly imagined that the infamous dunghill served as a sacrificial mound for the Jews in their worship of the devil.

According to a nineteenth-century (A.D.) traveler to the Holy Land, rabbinical Jews identified the privy as a house for unclean spirits, such as a dung-eating god. Care had to be taken when using the "house of honor" not to inhale, as the spirit could enter the body and cause disease.

Ancient Greece

The Greeks may have been original thinkers in politics and philosophy, but the worldly concerns of city sanitation were not a priority. Trade with other Mediterranean nations provided ancient Greece with valuable contacts. They undoubtedly assimilated elements from other cultures as well as spreading many of their own ideas. The Koros statues from the Hellenic period, for example, look suspiciously similar to Egyptian statues of gods and pharaohs. Sanitation systems were also similar in the two nations. Upper-class citizens of the Greek city Priene possessed latrines like those of high officials in Egypt.

However, the Athenian emphasis on beauty set them apart from their contemporaries. To the Greeks, visible beauty ruled society. Plutarch once remarked that the gods could forgive a sinner if he were a beautiful man. The idea of beauty even found its way to the chamber pot in the ancient Greek city of Sybaris. Its citizens had a reputation for extravagant luxury, laziness, and an ostentatious style. Because of their idleness, the Sybarites are thought to have invented the chamber pot. Apparently, leaving the room to relieve oneself was too much of an effort. So attached were the Sybarites to the decorative pots, they took them to parties or on their travels.

Etruscans

Settling in central Italy during the seventh century B.C., the Etruscans founded a nation strongly influenced by Greece. The rise of Rome overshadowed the Etruscans, who eventually assimilated into the Roman Empire. Despite their brief existence, they contributed significantly to the advancement of

Rome. Through the Etruscans, Greek philosophy and art became the basis of Roman civilization. And in the world of sanitation, the Etruscans left Rome the greatest sewer of the period: the Cloaca Maxima.

Six centuries before the birth of Christ, the Etruscans were building sewer trenches that drained into the Tiber River. Tarquinius Sperbus (534–510 B.C.), who ruled Etruria during the height of its short-lived realm, built the largest sewer of ancient times. The Cloaca Maxima measured over sixteen feet wide, and then was expanded by the Romans. Pliny (23–79 A.D.), the Roman scholar, described the Cloaca Maxima as "the most noteworthy achievement" of Rome. Seven branches of the sewer traveled under the city's streets to join the main channel, the Cloaca Maxima. During storms, the sewers were cleaned by the pounding force of the moving water. The Cloaca Maxima remains in use in modern Rome twenty-five hundred years after it was built.

According to Pliny there was dissension among the laborers (probably slaves) employed in constructing the sewer system. The project lasted many years and was arduous work. To avoid a life of servitude in building the sewer, a large number of the laborers committed suicide. The trend of suicide among his laborers alarmed Tarquinius. In response, he ordered that any worker who died by his own hand would have his body crucified and displayed in public. Denied burial, the body would be left to the birds of prey and scavengers. The shame associated with such a punishment was enough to reduce the number of suicides among laborers.

Romans

Crisscrossing the city, Roman sewers emptied into the massive Cloaca Maxima before entering the Tiber River. Convicts were forced to clean the sewers. But not all citizens enjoyed the benefits of the sanitation system. Only a few privileged homes could apply for a permit that entitled them to be connected to city sewers. Permits sold by Roman officials were costly and thus limited to rich homeowners.

The value placed on Roman sewers is obvious from the limited access given the public. The famous statesman Agrippa (63–12 B.C.) best expressed the fascination Romans carried for their sewer system. The Cloaca Maxima, emptying the neighborhoods of the Forum and the Aventine, was reportedly large enough to accommodate a wagon loaded with hay. In 33 B.C., while the sewer was being cleaned, Agrippa decided to test its limits. As Aedile of Rome, he traveled the entire length of the mighty sewer in a boat maneuvered by oars, to inspect the cleaning operation.

The presence of sewers affected the lives of common people in a more direct manner. Life and death occurred in the shadows of the extensive sewer system, and it became a popular place for Roman women to leave unwanted babies. Babies that did survive were rescued by barren women. These women would take the baby home and present the child as their own to their husbands.

The theme of death and the sewer continued in Rome's prisons. The underground Mamertine Prison had easy access to the Cloaca Maxima. Prisoners could be tortured, executed, and easily thrown into the sewer.

Vercingetorix, the leader of Gaul, returned to Rome with Julius Caesar after being defeated in 52 B.C. Julius Caesar exhibited Vercingetorix at his triumph six years later. The Gaul was then executed and presumably thrown into the sewer system.

St. Peter worked one of his many miracles while being held in the Mamertine prison. Having no water to baptize his fellow prisoners, he is said to have caused a spring to bubble up in the cell from the sewer.

Roman sewers alone did not control the accumulation of human excrement. Chamber pots, cesspits, and latrines contributed to a relatively clean city. Vases, called gastra, were deposited along the sides of roads and streets for travelers to relieve themselves.

Suetonius and other ancient Roman writers chronicled the strange lives of the emperors. Nero's insanity, Caligula's indulgences, and Tiberius's cruelty are well documented. Yet few have heard of the antics of Commodus (161–192 A.D.). Commodus is believed to have designed the piece of furniture bearing his name. Considering his preoccupation with bodily fluids, it is easy to believe he would create a chest to hold the implements. In *Scatologic Rites*, Captain John G. Bourke reported that Commodus was known to eat human excrement.

Another little-known Roman emperor joined the list of prominent people who died while in the privy. Heliogabalus, also known as Elagabalus, ruled Rome from 204 to 222 A.D. His death occurred while he was in the privy, and his body was then thrown into the pit below.

Roman citizens who could not afford to be connected to the sewer depended on public latrines as a comfortable source of relief. Ranging from the simple to the luxurious, public latrines were as much a way of life as the famous Roman baths. Romans were sociable people. For a fee, men could gather in the public latrines to attend to nature and gossip with neighbors. Parties were planned, politics discussed, and business deals made while sitting on the john. By 315 A.D. there were said to be more than 140 public latrines in the city.

The remains of the ancient latrine in Ostia outside of Rome misrepresent the once lavishly decorated building. Two thousand years ago, this public latrine possessed marble seats with beautifully carved dolphins separating the seats and providing some privacy for the patrons. Mosaics adorned the floors, depicting scenes of Roman life. To discourage graffiti, the walls were painted with images of gods and goddesses. Defacing the gods was considered a serious offense in Roman law.

The elegance of public latrines did not include the sanitation facilities. Trenches in front of the seats contained running water and the excrement of the patrons upstream. Buckets placed in front of the latrines held sticks fitted with sponges on the end. Patrons used the sponges to wipe themselves before replacing them for the next man. A variation of the public latrine had the seat emptying into a channel below and the trench in front used only for dipping sponges.

Chamber pots, or simple clay jars, served as the primary sanitation device of the common man. After use, the jar would be emptied into a public cesspit or directly out the window into the streets. During the night, the cesspits would be emptied of their contents by city-paid sanitary workers and carted away in wagons.

Pretentious Romans bought vessels made of gold or other precious metals and gems to pamper their behinds. St. Clement of Alexandria cried out against the gluttony of the rich during the first century A.D. He called particular attention to those who possessed luxurious piss pots.

Cicero had his own opinions on the chamber pot industry. He questioned the taste of connoisseurs who claimed the abil-

Ruins of public lavatories at Ostia, Italy (Julie Horan)

ity to tell in which workshop in Corinth a pot was made by its smell.

The Roman practice of throwing the contents of urine jars out of the window was to continue for centuries. Many people fell victim to intentional or unintentional deluges from above. The victim could take the accused to court for damages. Damages collected included medical expenses and wages missed due to current and future absences from work. As the culprit could not always be identified, fines were levied against all tenants living in the area from where the contents fell.

The chaos caused by chamber pots being emptied from apartment windows found expression in contemporary writings. The third satire written by Juvenal described the common incident as "Clattering the storm descends from heights unknown." John Dryden (1671–1700) completed the translation in the following verse:

> 'Tis want of sense to sup abroad too late
> Unless thou first hast settled thy estate;
> As many fates attend thy steps to meet
> As there are waking windows in the street:
> Bless the good gods and think thy chance is rare
> To have a piss-pot only for thy share.

Scatologic Rites

One word to describe ancient Romans would be *practical.* Sentiment and manners did not prevent them from using materials to their best ability. Juvenal was indignant with the industrious Roman merchants who found a way to make money off the need to pee. Fullers were merchants who dyed and fitted tunics for the populace, operating much like the modern tailor or dry cleaner. Human urine proved to be effective at removing grease and acted as a cheap source of dye. Human urine is composed of 98 percent water and 2 percent urea, calcium, phosphates, sodium, and ammonium. Ammonium breaks down to ammonia, the active ingredient. To ensure a steady supply of urine, fullers would place jars outside their establishment for use by the general public free of charge.

The enterprising Emperor Vespasian (9–79 A.D.) was constantly searching for a means to raise revenue. He charged fees for stewardship and other positions or titles. When he was informed that the senate had voted to erect a statue in his honor, Vespasian replied he would be content with the money raised to build the statue instead. Suetonius, the Roman historian, reported the extremes the emperor employed in obtaining money. He observed, "Titus [Vespasian's son] complained of the tax that Vespasian had imposed on the contents of the city urinals. Vespasian handed him a coin that had been part of the first day's proceeds: 'Does it smell bad?' he asked. And when Titus said 'No,' he went on: 'Yet it comes from urine.' " A different report of Vespasian's response was, "So we get the chinks, we will bear with the stinks." Since the average person produces two to three pints of urine a day, revenue raised from the urine tax could support a small army.

Romans who were too poor, or too drunk, to find a pot commonly used the staircase in the *insula*, or apartment building, as a latrine. The offender would simply squat in a corner while no one was around. Understandably, the practice was not welcomed by the other tenants. Hence, the origin of the expression "I didn't have a pot to piss in!"

Another famous Roman event was the banquet. Roman banquets were notorious for their immense amount of food and lengthy duration. Supper could last from two to as long as sixteen hours. For serious diners, rooms or pots were available to "barf up" the contents of their stomach and make space for more food. Eating for so long a time caused the bowels to move often. If a lounging Roman had need of a chamber pot, he merely snapped his fingers and a slave would conveniently deliver one.

The ancient Romans believed gods and goddesses ruled the forces of nature. Each god controlled a dimension of the earth and possessed a human personality. Demeter was goddess of the earth; Hades, god of the underworld; and Venus, goddess of love. Omitted from most mythology are the gods of human excrement. Cloacina was goddess of the common sewer. She received prayers when the sewer backed up or overflowed. Titus Tacius, an early Roman ruler, believed in the powers of Cloacina. He built a statue of her in a fitting temple, his privy. Some believed she shared the spotlight with the goddess of love. A statue of Venus, called Venus Cloacina, was found in a sewer.

Stercutius was the god of ordure or dung. Farmers found Stercutius important when fertilizing their fields with manure. Sir John Harington, the godson of Queen Elizabeth I and inventor of the first toilet with movable parts, believed Stercutius was really Saturn in disguise. Saturn was the god of agriculture, which would explain his close relationship with Stercutius. Crepitus was the god of convenience and by some accounts the god of flatulence. He received special recognition when a person had diarrhea or constipation.

Believing the gods and goddesses resembled man with all his desires and vices meant the gods also had bodily functions. Rain was viewed as the gods urinating.

Manners were important to the Romans despite the often foul nature of their environment. Commit No Nuisance was a command posted around temples to warn off persons who might offend the gods by relieving themselves in the vicinity. Deterrence of pissing on a statue came from the belief that the wrath of Jupiter, the chief Roman god, would be brought on the misbehaving soul.

Likewise, Romans avoided offending their emperors. A man consumed with a fit of colic refrained from breaking wind in the presence of Emperor Claudius. According to the man's physicians, his good manners contributed to his death. On hearing the cause of his death, the considerate emperor issued a proclamation that anyone in a fit of cholic was free to break wind in his company.

At its height, the Roman Empire spread as far as modern Britain. Traveling Romans built replicas of their cities in the new colonies. Ruins in Britain are similar in design to ruins found in Pompeii. Houses were built around a courtyard with service rooms such as kitchens and latrines placed in the corners of the house. Military posts in the colonies were protected by a fortress wall to deter the barbaric locals. The Romans were practical engineers, and latrines dotted the top of the walls. A Roman wall in Northumberland could accommodate twenty men using the latrines at once. Even Hadrian's Wall, built in 120 A.D. to keep out the Scots, had latrines along its massive length.

The exploits of the successful Roman army continue to be studied by modern military experts. But where the campaigns were devised is overlooked in the history books. Roman generals were notorious for drawing up battle plans while sitting on the pot. Later, this practice of conducting business while on the toilet was dubbed "French courtesies" after the seventeenth-century French court that embraced the custom.

Civilization began with strict attention to the disposal of human waste and the development of the "toilet." Building on the plumbing knowledge of the Etruscans before them, the Romans reached the height of their empire at the same time they reached the pinnacle of toilet evolution. As Reginald Reynolds points out in *Cleanliness and Godliness*, neglect of sanitation and reverence for the god of dung and goddess of the sewer contributed to the fall of the empire. The cavalier treatment received by the gods of excrement brought their wrath on Roman society. Consequently, the Roman Empire fell. It is ironic that the final blow to the empire came from the barbaric Germanic tribes of the north, who, though not as sophisticated as the Romans, regarded dung as an important tool for agricultural use and as building material.

Savages

As the Romans spread beyond their borders into lands of barbarians, they encountered people whose approach to living greatly differed from theirs. The sanitation systems of the "uncivilized" world must have appalled the genteel Romans.

Germans

During the third century, the Celts of the British Isles used pits around their huts for garbage, animal carcasses, and human waste. As many as 360 could be found in a settlement. But the Germanic tribes were more crafty. They dug pits for storage and placed manure and garbage on top. The manure dissuaded invaders from searching for valuables. Some pits were even lined with limestone.

Vikings

The Romans found competition for England from the Vikings. After years of hit-and-run plundering, the Vikings settled on the British Isles. By 71 A.D. York, north of London,

Viking in a privy (Jorvik Viking Centre, York, England)

boasted a large population of Vikings. Recent excavations of the settlement have revealed details of the Viking way of life.

The Viking diet consisted of grains, fish, and berries. Samples of human feces suggest a tormented existence. Vikings suffered from whipworm, roundworm, and maw worm. The samples were distributed haphazardly about the settlement, inside and outside houses, suggesting that squatting anywhere was acceptable. Latrine pits located near wells contributed to the constant appearance of disease. The most advanced facility discovered resembled a sewer from which small branches conducted waste material to the Floss river in York. Whether it was a sewer or just a ditch is unclear.

The Jorvik Viking Centre in York, England, documents the Viking existence through lifelike displays of wax figures. The curators feared that the photograph of a Viking squatting on a latrine holding a moss for wiping (p.19) might offend viewers. A screen of tree branches was therefore placed around the latrine to cover the offensive sight.

Few people grow up wanting to study ancient excrement. But Andrew Jones, a paleoscatologist at the University of York, has been very successful in tracking down old poop. His archeological excavations in York have unearthed a one-thousand-year-old piece of excrement left behind by a Viking settler. Affectionately named "Lloyds Bank Turd" after its location under Lloyds Bank, the piece of excrement is insured for $34,000.

Ancient Ireland, stark and barren, offered little protection for its inhabitants from the weather or invading marauders. During the first century the Irish built ring forts to provide protection from invading Vikings and enhance their communal environment. The ring forts consisted of round earthen walls containing several huts. Excavations have identified latrine pits lined with poles made from branches and reeds. Lining cesspits prevented or delayed seepage of waste and contamination of ground water. Some sort of wood structure covered the pit. Presumably it offered shelter and some comfort.

⚜ **2** ⚜

The Middle Ages:
Sir Lancelot's Toilet

The legends of King Arthur, Sir Lancelot, and Robin Hood depict a romantic era of innocence and chivalry. Knights vied for the privilege of carrying the symbol of a chosen lady into tournament. Lords and ladies wandered the halls of forbidding castles, casting shadows along candlelit passages. The Middle Ages appear to be the romantic golden years of Western civilization.

A more accurate picture of the medieval times, dating from 500–1500 A.D., would be of shacks leaning against castle walls, housing pock-marked townspeople. The muddy streets were home to animal carcasses, stagnant water, rubbish, and human excrement. Wars were common as lords tried to gain control of the different regions.

But the turmoil caused by warfare was minimal compared to the destruction produced by disease. A series of epidemics occurred during the years 1348–1350. The plague known as Black Death was fueled by the rampant filth flowing in the streets, the lack of personal sanitation, and the crowded conditions of medieval towns. Within a few years, one-third of the entire population of Europe died. The foul smell of death and disease led many people to carry flower petals in their pockets hoping to ward off the stench. Children created playful songs describing the grotesque environment. "Ring around the rosie, pocketful of posies, ashes, ashes, we all fall down" illustrates the progression of disease. The victim would break out in a fever causing rosy cheeks. Flowers would be placed on the sick person to counter the smell. Finally after turning an ashen pale color, the victim died.

Understanding medieval times requires examining the sanitation systems in castles, monasteries, and town homes. Only through studying waste disposal can the answer be found to why disease nearly wiped out an entire continent.

The Environment

Authors of the period left a record of the living conditions in many of Europe's major cities. Their descriptions provide an image of unbelievable squalor and stench. In the city of Nuremberg, Germany, for example, open sewers crossed the distance from each house to the river. When the river could no longer manage the disposal of so much waste, it was carted outside the city. Low tides were a particularly nasty sight as the shortage of water prevented the sewage from floating away.

At times the odors associated with daily living during the Middle Ages were unbearable. Perfume and flower petals helped alleviate some of the smells. Other methods in aiding the nose included the burning of frankincense, popular with the monks, and carrying pompadours. Henry VIII was known to carry a pompadour made of an orange skin stuffed with potpourri into his privy.

Medieval Paris was a typical city of its period. A huge wall provided protection from outside attackers. Human waste collected in Paris was dumped outside the city walls and gave some relief inside from filth. Unfortunately, as Paris grew, so did the dung pile. Eventually, the dung pile became so large that the city wall had to be extended for security reasons: enemies could attack the city from the top of the dung heap.

To contemporaries, "shooting the bridge" referred to traveling by boat beneath London Bridge during high tide. Misjudging the river could result in the boat being hurled against the bridge. Added to the perils of traversing the river were the public latrines serving the 138 houses built along London Bridge. Waste from the bridge latrines dropped directly into the Thames River, joining the filth dumped upstream by other citizens. Persons journeying on the footpath below the bridge or by boat were open targets for the remains of someone's lunch. Hence, the popular saying that the bridge was "built for wise men to go over and fools to go under."

A convenient feature of the public latrines on London Bridge was the front and rear entrances. A man evading his creditors took advantage of the latrines to make an escape.

Dirty Dungeons

As the concept of privacy was just beginning to evolve at this time, castles had few partitioned rooms. The lord of the manor and his family had separate living quarters, but the rest of the household shared a large communal room. But life in a castle offered more of a homy atmosphere than a ruin suggests. Tapestries hung on the walls and rugs covered the floors, furniture was sparse but sturdy, and restrooms, known as garderobes, were even available.

Garderobes were small seats built into the recess of castle walls and located in towers or near banquet halls. They measured as little as three feet wide and were approached by a right-angled wall. The Tower of London's privy is located next to the banquet hall where Henry VIII dined. However, it is doubtful that the rotund Henry was able to fit in the tiny garderobe.

Garderobe on unknown castle in Switzerland (Karn Wong)

The term *garderobe* means clothes closet since it resembled a dressing room. Although garderobe is generally used to describe the castle's privy, the small room was identified by various names in the Middle Ages. During the sixteenth century, the vault was known as the "priest's hole" because it was used to hide Roman Catholic priests from persecution in England. The majority of names, however, were euphemistic terms designed to hide the purpose of the room; they included: "the place of easement," "the oratory," and "the chapel." In addition, the term *privy* became popular when the garderobe was recognized as a place of solitude, ideal for reading a good book. "Privy" takes its name from the Latin word for privacy.

Often built near fireplaces or kitchen flues, garderobes caught the heat rising from the kitchen fires below to warm their cold, stone seats. Waste coming from the garderobe dropped several hundred feet into the moat below, thus preventing the problem of "back splash." Unfortunately, after years of receiving the castle's waste, the odors arising from the moat made life within the castle walls unpleasant.

As ruler of England, King Henry III worried about toilets: his own, of course. In 1313, tired of the ugly brown stain on the walls outside the garderobes, Henry ordered the constable of the Tower of London to build a hollow column, or flue, to hide the offense. At the same time, the king asked for several privies to be built in his other houses. He even built an underground drainage system for waste water at Westminster Palace.

The Count's Castle (also called Gravensteen) in Ghent, Belgium, offers an excellent example of castle garderobes. In the living quarters, there is a garderobe that accommodates one person. On the fortress wall, the garderobe served the less-privileged occupants of the castle. As privacy was a luxury for the entitled, the privy on the wall held two people at a time. The offerings of both facilities flowed into the river running adjacent to the castle walls. Using the garderobe on the castle wall must have been a horrible experience during the freezing temperatures of winter, causing many bottoms to freeze to the stone seat.

Medieval castles were not the impregnable fortresses they appeared to be. They possessed a weak link in the form of the garderobe. During times of siege, the enemy could climb the walls of the castle and gain entrance through the garderobe's hole, a harrowing experience for any soldier. Additionally, sharp shooters took advantage of the opening during their assaults with arrows.

Another means of invading the castle was to wade through the muck of the surrounding moat. As it turned out, using the moat as a cesspit strengthened the defense of the castle. Few soldiers would swim through a moat full of dung to reach the castle. Only on rare occasions, for the comfort of the castle's inhabitants, were the moats cleaned, and the dung carted away.

Many castles contained underground passages that connected the fortress to a nearby river. These passages linked several of its rooms. Intended for use as a sewer or a tunnel for waste water, the underground labyrinth often served as a

Garderobe at Count's Castle in Ghent, Belgium
(Julie Horan)

route for secret meetings or an escape when necessary. For example, Rochester Castle in England owned an underground sewer that passed the dungeon in its path to a nearby river. When the opportunity arose, many prisoners escaped the prison in this manner.

The Meanderings of Monks

Living accommodations for monks in medieval times were vastly superior to those of the majority of townspeople and in many circumstances those of lords and ladies. The secluded nature of the monastery and the monks' disciplined nature concerning human waste enabled a number of them to escape the ravages of the Black Death plague.

Although monks did not condone frequent bathing except in times of illness (bathing, especially with hot water, was believed to stimulate the body), they valued efficient latrines and lavatories. Monasteries were built next to flowing streams, allowing the waste produced by the inhabitants to drift downstream. Latrines were located behind the dormitories, usually accessed by a bridge. The Canterbury Christ Church monastery built in the twelfth century contained underground pipes, cisterns, baths, and privies. To please the church, the privy seats were separated by a paneled wall, for the sight of a naked body caused arousal and temptation to the soul.

The Tintern Abbey in England built their "necessary house" too close to the Severn River. The benefit of the location was that during high tides the river cleaned the latrines. But during floods, the entire building could be swept away, and waste would back up or explode in the privies.

—エ—

The powers of heaven and earth were placed in the hands of church leaders during the medieval era. Cardinals and bishops controlled the lives of their subjects by handing down the word of God, often to their own benefit. The followers of Christianity readily accepted church penalties in an effort to win the grace of God. Occasionally the leaders themselves worried about their own souls and asked for prayers from others.

The Bishop of Wells used his influence with the local townspeople to save his soul. Graciously, he allowed the local town to divert a portion of the river passing the monastery for the citizens to dump their human waste. In return for the bishop's benevolence, the townspeople were required to say a prayer once a year for his health.

-I-

For the most part, monks protected their privacy. They were reluctant to have the outside world interfere with their daily lives. Saint Hugh of the Cluny monastery in England was an exception. During the eleventh century, Saint Hugh built an addition to the monastery to accommodate visitors and travelers of repute. The building contained beds for forty men and thirty women. Interestingly, each accommodation included an individual latrine. To avoid contaminating all the monks, contact with the outsiders was restricted to one monk who managed the hostel and emptied the latrines.

-I-

Medieval monasteries were the bastions of intellect, more by exclusion than for any other reason. Only the men of God and the aristocracy were allowed to learn Latin and study time-honored writings. The secluded lives of monks gave rise to a reputation of mystery and power. The reality of life in a monastery was hardly as romantic.

Archaeologists have discovered much about the lifestyle of medieval monks by studying their cesspits and garbage dumps. The monastery privy pits have produced torn fragments of scratchy, old wool cloth used to wipe the monks' bums. Also found in the pit were buckthorn seeds, a common remedy for constipation. Days spent kneeling in prayer and studying the works of antiquity took their toll on the monks' bowels.

-I-

Theologian Martin Luther (1483–1546) represents the thoughtful, anguished soul of a man of God. Luther's famous *Ninety Theses* criticizing the Roman Catholic church did not come to him in the form of a miraculous thunderbolt. Rather, Luther's contemplations on the role of the church were slow and tortuous in coming. Frequently stricken with constipation, Luther had plenty of time to deliberate on the evils of the church.

God's call for service came early in Luther's life, as did his bowel trouble. According to his own writings, Luther first considered becoming a monk while sitting on the "necessary" defecating. In this holiest of sanctuaries, Martin thought about his father's request that he marry a young lady of wealthy means. Suddenly Luther cried out, "Help me St. Anne . . . I want to became a monk." However, after the initial pain passed, he had second thoughts. In the end, Luther avoided marriage and joined the church.

-I-

In the fifteenth century, travelers to the Holy Land had to learn to survive for months on the open sea. The necessities of life could not be ignored: food, water, and "going to the bathroom." A Dominican monk, Felix Faber of Ulm, traveled as a pilgrim from Europe to the holy lands in 1480 and 1483. He gives an astonishing portrayal of life on the open sea in *A History of a Private Life.*

> As the poet says, "A ripe turd is an unbearable burden." A few words on the manner of urinating and shitting on a boat. Each pilgrim has near his bed a urinal—a vessel of terra-cotta, a small bottle—into which he urinates and vomits. But since the quarters are cramped for the number of people, and dark besides, and since there is much coming and going, it is seldom that these vessels are not overturned before dawn. Quite regularly in fact, driven by a pressing urge that obliges him to get up, some clumsy fellow will knock over five or six urinals in passing, giving rise to an intolerable stench.
>
> In the morning, when the pilgrims get up and their stomachs ask for grace, they climb the bridge and head

for the prow, where on either side of the spit privies have been provided. Sometimes as many as thirteen people or more will line up for a turn at the seat, and when someone takes too long it is not embarrassment but irritation that is expressed. I would compare the wait to that which people must endure when they confess during Lent, when they are forced to stand and become irritated at the interminable confessions and await their turn in a foul mood.

At night, it is a difficult business to approach the privies owing to the huge number of people lying or sleeping on the decks from one end of the galley to the other. Anyone who wants to go must climb over more than forty people, stepping on them as he goes; with every step he risks kicking a fellow passenger or falling on top of a sleeping body. If he bumps into someone along the way, insults fly. Those without fear or vertigo can climb up to the prow along the ship's gunwales, pushing themselves along from rope to rope, which I often did despite the risk and the danger. By climbing out the hatches to the oars, one can slide along in a sitting position from oar to oar, but this is not for the faint of heart, for straddling the oars is dangerous, and even the sailors do not like it.

But the difficulties become really serious in bad weather, when the privies are constantly inundated by waves and the oars are shipped and laid across the benches. To go to the seat in the middle of a storm is thus to risk being completely soaked, so that many passengers remove their clothing and go stark naked. But in this, modesty suffers greatly, which only stirs the shameful parts even more. Those who do not wish to be seen this way go squat in other places, which they soil, causing tempers to flare and fights to break out, discrediting even honorable people. Some even fill their vessels near their beds, which is disgusting and poisons the neighbors and can be tolerated only in invalids, who cannot be blamed:

a few words are not enough to recount what I was forced to endure on account of a sick bedmate.

Faber continues his memoirs with advice for those following in his footsteps on a visit to the holy lands. He specifically highlights measures to be taken in the maintenance of one's bowels.

The pilgrim must be careful not to hold back on account of false modesty and not relieve the stomach; to do so is most harmful to the traveler. At sea it is easy to become constipated. Here is good advice for the pilgrim: go to the privies three or four times every day, even when there is no natural urge, in order to promote evacuation by discreet efforts; and do not lose hope if nothing comes on the third or fourth try. Go often, loosen your belt, untie all the knots of your clothes over chest and stomach, and evacuation will occur even if your intestines are filled with stones. This advice was given me by an old sailor once when I had been terribly constipated for several days. At sea, moreover, it is not safe to use pills or suppositories, because to purge oneself too much can cause worse trouble than constipation.

The Disposal of Dung

Disposal of human waste in medieval cities occurred either by dumping it in the river, burying it in pits, or shipping it out of town. Convenience ruled the day, not health concerns.

Rivers in England swept the dung away until feces filled the depths of several rivers, causing stagnation. Fleet River in London collected the remains of eleven public latrines on a bridge and three sewers. Not surprisingly, the river quit flowing, and Fleet River became Fleet Street. But before that advent, the smell of the river became so unbearable that the monks of White Friars complained to Parliament, blaming the atmosphere for the deaths of several monks. Even the scent of frankincense burning on the church alters could not overcome the stench of dung.

The dike at Fleet Prison was in no better condition. In 1355, Edward III ordered an inquiry into the buildup of dung near the prison. The pile proved deep enough to float a boat. But Parliament had its own problems with dung-infested waters. Latrines from London Bridge dumped two thousand tons of excrement a year into the Thames River. As Parliament convened in a building by the river, hot summer days with no breeze incapacitated the law makers. Sheets of pressed rose petals were placed in the windows of meeting halls to alleviate some of the smell.

If no river or moat was handy for dumping, a cesspit or the open ground served the purpose of waste repository. While city populations remained low, much of their human waste could be collected and sent to farms in the countryside as fertilizer. As the cities became crowded, residents dumped their waste into city streets. Eventually, the offensive material blocked several roads. Sherborne Lane became known as "Shiteburn Lane." Pedestrians were prevented from crossing Ebbegate Road because of the pile of dung blocking the passage.

Recurring epidemics encouraged many European officials to require that people use cesspits, not rivers, as a means of disposal. Unfortunately, their warnings often went unheeded. Parisian police issued orders for the installation and use of drains and latrines during the years 1522, 1525, and 1539. Not convinced of the imminent need, the people of Paris continued to dump their waste into the streets of the city, making it the "city of smells" rather than the "city of lights."

In England, officials faced problems with citizens who did not empty their cesspits. In 1328, the Mere brothers, William and Adam, neglected to care for their cesspit. Eventually, dung began to seep through the cesspit into their neighbor's walls. The Mere brothers paid a hefty fine for their poor housekeeping skills.

One of the best paying jobs in medieval England was that of the *gongfermor*, or raker. A gongfermor worked during the night cleaning the cities cesspits. The name gongfermor comes from the Saxon word *gang* meaning to go off, and *fermor* from *fey*, to cleanse. City councils hired the gongfermors when the stench from dung-filled pits offended even the resilient medieval nose. In addition, the gongfermors sold the contents of the cesspits to farmers in the countryside as fertilizer. The unenviable job entailed shoveling the contents into barrels and disposing of them outside the city. The sound of the "night wagon" rambling through the streets continued to be heard for centuries to come.

Cleaning privies was dangerous work, as Richard the Raker of London discovered. A gongfermor, Richard met his death in 1326 in one. He fell through a rotting board into the cesspit below, drowning in the muck. Ironically, Richard the Raker died in his own privy.

Jerrys, Chimneys, and Thrones

"One taught an excellent rule to keepe a chimney from smoking, and a privie from stinking, viz. to make your fire in your privie, and to set the close stoole in the chimney."

The Metamorphosis of Ajax

Not all people living in the Middle Ages enjoyed the luxury of garderobes or latrines for the call of nature. The most common manner of relief required its user to squat over a pot. Early in the period, chamber pots were made from earthenware or tin in a crude design, usually with a wide base and opening and a narrow neck. Few of them bore decorations, making it difficult for historians to determine whether a pot was intended for cooking or sitting. Medieval residents referred to their chamber pots as "originals" or "jerrys."

By the sixteenth century, the discovery of Chinese porcelain led to highly decorative chamber pots. Ironically, chamber pots were displayed openly in the early part of the medieval period, but as the pots became more decorative, the users began to hide them in furniture.

The urinal, an elongated, narrow version of the chamber pot, held obvious advantages for men. Doctors in the Middle Ages diagnosed disease by examining urine through glass urinals, a process known as uroscopy. Although not exactly scientific, the examination of urine was believed to help physicians identify illness. Doctors would look for discoloration of the urine, signifying illness. The cure remained the same— apply leeches to the offending region.

Toward the end of the Middle Ages, close-stools became popular among the rich. The close-stool consisted of a chamber pot placed inside of a wood box, enabling the user to sit instead of squat while defecating. Louis XI of France (1423–1483), a pious, quiet man, preferred to spend his hours in prayer rather than attend to the worldly demands of managing a country. Louis owned a special close-stool that gave him the much desired privacy. The close-stool had a curtain hanging on an iron framework that enclosed him and the box when in use. In addition, the saintly king used herbs to freshen his royal air.

A few ingenuous individuals devised alternative devices for the collection and disposal of feces. The consequences of their inventions were often messy and futile. In Paris during the fifteenth century, some town houses shared a privy that connected the two buildings on the second floor. Contents from the privy fell to the ground below. A few of these privies were merely a couple of planks suspended from the second level of the house and supported by poles. The story of Andreuccio illustrates the danger that stems from such homemade privies.

The danger of a medieval French latrine

As a young man, Andreuccio broke through some rotten boards in the second-floor privy while making use of the facilities. He suddenly fell several feet into a pile of dung. Not seeing the need to clean himself off, Andreuccio hurried to a rendezvous with friends. Although Andreuccio was not bothered by his smell, his friends could not abide the stench and dipped him into a drinking well to clean him off.

Finally a fledgling, but honest, attempt was made at inventing a working toilet. In 1449, Thomas Brightfield built a stone toilet flushed by rainwater through a pipe that emptied into a cistern. Unhappily, the invention did not include the important feature of a valve to prevent backflow and unpleasant "aromas." Brightfield was before his time.

Home improvements during the Middle Ages often ended in mayhem. In 1321, a woman from London decided to improve the design of her home. She built an addition with a privy connected by a pipe to the rain gutter. The gutter soon became clogged, creating a smelly mess for the surrounding neighbors. The woman received a heavy fine from the local magistrates for her efforts. In 1347, two men chose a devious means of ridding themselves of collected waste. They diverted the pipe carrying their dung into their neighbor's cellar. The crime was quickly noticed and the men arrested.

Advice to Be Heeded

The Medieval Privy

Beware of draughty privys and of pyssyge in draughts,
And permyt no common pyssynge place about the house
And let the common house of easement to be over some water
Or else elongated from the house.
Beware of emptying pysse pottes and pyssing in chymnes.

Andrew Boorde (1500–1549)
English physician and writer

Privy Matters

Lusty, barbarous, and plundering men marked the beginning of the British throne in the early medieval years. One king in particular made his presence known early in life. According to Henry of Huntington (d. 1154), author of the chronicle *History of England*, "Ethelred, son of King Edgar, and brother of Edward, was consecrated king before all the nobles of England at Kingston. An evil omen, as St. Dunstan interpreted it, had happened to him in his infancy. For at his baptism he made water [urinated] in the font; whence the man of God predicted the slaughter of the English people that would take place in his time."

Written after the king's death, Henry of Huntington's remarks definitely had the advantage of hindsight. Ethelred the

Unready, as Ethelred was known (979–1016), proved a disastrous king. His forces were routed by the invading Danes in 1002, the St. Brice's Day Massacre.

-I-

History made famous the tale of Thomas Beckett's murder in the cathedral. Edmund Ironside's murder in the loo is significantly less well known. Son of the ignominious Ethelred the Unready, Edmund Ironside, a courageous man, ruled for only one year. Edmund found himself continually battling Cnut, a Dane, for control of the realm of England. This great fighter died at the hands of his enemies while sitting on the "pot." Again, Henry of Huntington relates the tale:

> King Edmund was treasonably slain.... Thus it happened: one night, this great and good king having occasion to retire to "the house for relieving the calls of nature," the son of the ealdorman Edric, by his father's contrivance, concealed himself in the pit, and stabbed the king twice from beneath with a sharp dagger, and leaving the weapon fixed in his bowels, made his escape. Edric then presented himself to Canute [Cnut], and saluted him, saying "Hail! though who art sole king of England!" Having explained what had taken place, Canute replied, "For this deed I will exalt you, as it merits, higher than all the nobles of England." He then commanded that Edric should be decapitated and his head placed upon a pole on the highest battlement of the Tower of London. Thus perished King Edmund Ironside, after a short reign of one year, and he was buried at Glastonbury, near his grandfather Edgar.

-I-

A man living in fifteenth-century London had a fight with his wife, who hit him over the head with the chamber pot. The man left the house to sit on the curb and contemplate their argument. As he sat thinking, his wife poured the contents of another chamber pot out the window and onto his head. The husband was heard to reply, "It seems we will get rain tonight." Evidently, using the chamber pot as a weapon was fairly common. In 1418, a man named Baudet was banished from Paris for breaking one over the head of a lady.

-I-

As men left home during the Crusades, they worried about the chastity of their women. To ensure loyalty, many Crusaders from England and France used chastity belts to lock up their wives' lower body. Small holes in the chastity belts worn by the women allowed the release of bodily fluids. Keeping the device clean proved difficult if not impossible. The smell became unbearable for some women, who committed suicide rather than endure the discomfort of the chastity belt.

-I-

A bizarre rite connected with the consecrating of a new pope had the newly elected holy man sit on a chair designed with an opening to hold a chamber pot. Called the "Sedes Stercoraria," or fecal throne, the pope's chair held no pot. Instead, the youngest member of the church present had the dubious honor of crawling under the chair, reaching through the hole, and touching the pope's genitals. Bartolomeo Platina believed the practice of using a chair in such a manner was meant to remind the pope of his nature as a man not a God.

In reality, the rite was a means of checking the pope's sex. It seems that Pope John VIII was actually a woman, Joan. Her true sex was discovered when she died while giving birth. Aware of the famous Pope Joan, Peter the Great of Russia used a similar rite as part of a diversion acted out by his drunken companions. Pretending to be cardinals, Peter and his friends appointed an unlucky man, Buturlin, pope. While Buturlin sat on the special chair, Peter reached under and grabbed Buturlin's genitals, crying "He has an opening!" Everyone was much amused.

-✠-

Monks viewed flatulence as the devil's temptation. Breaking wind supposedly increased the desire of a man for sex by placing pressure on his genitals. Monks were warned to avoid foods that would cause gas.

-✠-

Sexual harassment in the medieval age was difficult for women to address along acceptable lines. Pope Alexander IV (d. 1261) told the story of a woman harassed by her local priest who found a creative manner to revenge her attacker. During confession the roguish priest attempted to rape the woman. The woman escaped the priest by feigning a desire to meet with him later for a romantic rendezvous. Instead, she sent her abuser a pie to express her feelings. It was no ordinary pie. The pie was baked with the woman's excrement. As luck would have it, the priest gave the pie to his bishop as a token of respect. Needless to say, the priest's career in the church was short-lived.

-✠-

Regular bowels would seem to be a blessing in a time of limited facilities and questionable diet. However, King Bruce of Scotland (1274–1329) discovered the perils of a king maintaining a routine in any facet of daily life. The king rose early every morning to sit upon his privy. Knowing the king's routine, three of his enemies waited in his privy to kill him. Luckily, Bruce carried his sword with him on his privy excursions. The King of Scotland swiftly killed the intruders before settling down to the business at hand.

-I-

Privacy was paramount in the selection of a pope. What better place to achieve privacy than in the room named for its discretion—the privy. In the fifteenth century, the cardinals of Rome gathered in the latrine to choose Pope Pius II (1405–1464) as their leader.

-I-

In 1183, Holy Roman Emperor Frederick I lost the leaders of his kingdom in central Europe in an ironic tragedy. Eight princes and several knights were gathered in the great hall of Erfurt castle for a meeting of the diet. The weight of the attendees proved too much for the wood floor. Wood splintered and cracked, sending the entire group of nobles several feet into the cesspool beneath the castle. Everyone perished, drowning in the pit of excrement.

-I-

James I of Scotland (1394–1437) discovered the hard way the dangers of "pissing off" the noblemen. Unearthing Sir Robert Grames's plan to overthrow him, King James had the nobleman banished from Scotland and all his property confiscated. Grames vowed to kill the king. The opportunity for murder arose while King James visited Black Monastery in Perth. The evening followed the normal routine for the royal party. The king and queen entertained a group of aristocrats with dinner and discourse. After the royal entourage left for the evening, the queen and her ladies in waiting remained present during the king's preparation for bed. Suddenly the noise of a contingent of armed men filled the midnight air. Sir Robert Grames, with a group of noblemen also opposed to James, was planning to make good on his promise to kill the king.

James immediately realized he was doomed. Unfortunately, he could find no means of escape. The windows were welded shut with lead and would require several men to break open. The women in the chamber were frantically trying to lock the door while the king searched for a hiding place. Rejecting the obvious locations of under the bed or in the wardrobe, James picked up the poker for the chimney and decided to fight his assassins. Suddenly he remembered that connected to the chimney was a privy pit. He lifted the hinged door on the wooden floor and climbed into the disgusting pit. The pit possessed a flue that connected it to a ditch outside the monastery. The King believed he could fit through the flue and escape. Two problems faced James at this point. He was very fat, yet he was optimistic on his chances of fitting through the flue. Secondly, days before the attack, James had ordered the flue closed because he had lost a ball he was playing with. It was stuck in the privy. The best he could hope for was to hide till his assassins left.

Sir Robert Grames's men had no problem breaking down the door to the king's chamber. In the process, they seriously wounded many of the court ladies with their axes. Entering the room they found the queen standing numbly before the chimney. A nobleman attacked the queen and would have killed her if Grames had not stayed his hand. Grames wanted the king dead, not his wife.

Searching the chamber, the traitors could not find the king and withdrew to search other parts of the building. Not hearing any noise overhead, the king began to yell to the queen's women to lift him out of the pit. As the women tried to help the large king, one of the women fell into the pit with him. About that time, Grames's men returned to the chamber having remembered that the privy could serve as a hiding place. Finding James in the pit with a woman, they laughed at the sight. The king making love to his wife's maid in the privy! Next, one of the traitors jumped into the privy intent on killing the king. But James put up a good fight. He was a strong man. Another assassin joined the fight. Finally, Robert Grames lowered himself into the pit and with a large sword killed King James personally with sixteen stab wounds to the breast.

-¤-

According to Sir John Harington, the diabolical Richard III of England (1452–1485) was rumored to have planned the deaths of his two nephews while he sat in the privy. Richard acted as regent for his young nephews after the death of their father. When the princes "disappeared," Richard assumed the throne.

⊰ 3 ⊱

Premodern Europe: Raunchy Reveille

1500–1700

To keepe your houses sweete, cleanse privie vaults,
To keepe your soules as sweete, mend privie faults.

The Metamorphosis of Ajax

The Renaissance of fifteenth-century Italy, spreading north-ward, in the sixteenth and seventeenth centuries sparked the imaginations of painters such as Holbein and Brueghal. Martin Luther's criticism of the Catholic Church resulted in the creation of rival protestant churches and the reformation of the Catholic Church. Religion lost power and royalty gained. In terms of toilets, the changes meant that the former hygiene practices of the monastery monks was replaced by the flamboyantly decorated apparatuses of the rich. The habits and devices used by the royals were infinitely more entertaining than the pious cleanliness of the monks.

Though the common people of early modern Europe continued to rely on chamber pots and privies, innovators began experimenting with different methods for "going to the bathroom." It was during the sixteenth century that the first working toilet with moving parts was developed. Unfortunately, the majority of the populace found the toilet to be a vulgar device and refused to consider using it. Emerging from the pious Middle Ages, many societies during the Renaissance continued to view excessive attention to the body, such as bathing or elaborate means of evacuation, as unholy or merely uncouth. The most important addition to sanitation died an acrimonious death due to lack of interest. The practical reason for the failure of water closets during the sixteenth century was the scarcity of water systems to support the operations of the device. It was not to be revived until the Victorian era—nearly three hundred years later!

-I-

Leonardo Da Vinci left a blueprint containing plans for a sanitary city. Had anyone seriously considered Da Vinci's ideas, the sixteenth century might perhaps have been a pleasant time to live. Writing in his journals, Da Vinci considered the problems of overcrowded cities and proposed several solutions. His answer to waste was to provide an abundance of latrines. He described his creation, "The seat of the latrine should be able to swivel like the turnstile in a convent and return to its initial position by the use of a counterweight; and the ceiling should have many holes in it so that one would be able to breathe."

In addition, he proposed spiral staircases to deter the use of the stairwell as a toilet. Da Vinci described his city: "The privies, the stables and suchlike noisome places are emptied by underground passages, situated at a distance of 300 brachia from one arch to the next, each passage receiving its light through the openings in the streets above. And at every arch there should be a spiral staircase; it should be round because in the corners of square ones nuisances are apt to be committed. At the first turn there should be a door of entry into the privies and public urinals, and this staircase should enable one to descend from the high-level road to the low-level road. The site should be chosen near to the sea or some large river, in order that the impurities of the city which are moved by water be carried far away." (1490s)

-I-

Since no one listened to Da Vinci, chamber pots ruled the day. No longer made of crude tin or earthenware, the chamber pot reflected the status of the user. James I owned a pot made of silver. Cardinal Mazarin's was made of glass with velour and a gold band with silk and gold tassels. Louis XIV, true to his pompous nature, had a chamber pot made of gold decorated with his royal coat of arms. Louis owned spare pots used for both traveling and warfare. After dinner parties, when the ladies had retired to the salon, chamber pots were provided for the men in dining rooms so they need not leave the company of the other men.

-I-

Disposing of human waste as their ancestors did, seventeenth-century city dwellers often threw the contents out their windows into the streets below. Even though long used to disgusting and acrid smells, people could only endure so much. Henry VIII of England took time-out from complaining about his wives to complain about the town of Cambridge in 1544. He railed against the mounds of dung, filth, and mire lining the streets. He cited the condition of the city as the cause of health problems. Five years later, under the reign of his son, Edward VI, Parliament called for the building of sewers to accommodate the waste.

Apathy toward government regulations has long been a problem in Western civilization. A few sewers were built, but no one gave serious consideration to the problem. The

Parisian government ordered drains and latrines to be built and used by inhabitants in 1522, and again in 1525, and again in 1539. No action was taken.

In a custom retained from Roman days, residents of European cities disposed of the chamber-pot contents by throwing them out the window. Thankfully passersby were politely warned of the impending doom. A French woman cried, "*gardez l'eau*" prior to the contents falling. The English adapted the cry to "gardy-loo," possibly the precursor to the toilet being called the "loo." "Lord have mercy on you" was often added to the English cry. In Italy it was, "Take away your lantern." Men with good manners walked on the left side of women to shield them from the grotesque attack out the window. The custom survives today.

Some people viewed the chamber pot as disposable. Instead of dumping the contents of the pot, they dropped the pot into the street. The chaos caused by this violent means of disposal led authorities in Paris to ban throwing chamber pots out windows in 1395. However, French citizens continued the practice into the seventeenth century. The governor of Versailles issued a statement forbidding "all people from throwing human waste and other immodities out the window."

Who cleaned the mess that accumulated in the streets? If the residents were lucky, a "sanitary engineer" was hired for the job. During the Tudor reign in England, public health officers were elected. They were called scavengers and were responsible for cleaning the streets and resolving complaints. Their most notable contribution to history was the daily street cleaning during the plague of 1666. These men-of-the-night advertised their trade and passed out business cards.

Hearing complaints concerning the disposal of waste offered some of the best entertainment of the time. (Forget seeing *MacBeth* at Folger's Theater: the sanitation court was the place to be.) Citizens of London would take their complaints before a judge appointed to rule on sanitation offenses. Qualification for this judgeship required experience

"Night," from *Times of the Day*, William Hogarth (Huntington
Library, California)

as a scavenger. The majority of cases before the court were
brought by individuals accusing their neighbors of dumping
waste on private property.

One case involved an entire block of houses. The inhabitants of West Street in London dumped their dung into the

yard of a resident named John Davis. The Scavenger Court ordered the offenders to remove the filth before the Feast of St. Andrew. In addition, the magistrate felt it necessary to admonish those responsible for dumping excrement in the streets and the local churchyard.

The cure for all this public evacuation was the development of the privy. Slowly, hygiene habits began to include the privy as a part daily life. Wide use of the device, however, did not take hold until the late nineteenth century.

By the 1500s, the appearance of the privy signaled an important change in attitude toward privacy in high society. Early privies consisted of a closet with a close-stool or chamber pot in it. The privileged people who possessed them valued the privacy offered by closed doors. No longer were the sights and sounds of answering nature's call viewed by an audience. The aristocracy of Europe gradually incorporated the small room into their homes. France recorded the existence of privies in the castle of the duke of Burgundy and in the neighborhoods of Saint-Genevieve in Paris.

-I-

Determined to make a contribution to the world, Sir John Harington (1561–1612) introduced his invention—the Ajax—a "sweet-smelling" privy. Harington was a favorite godson of Queen Elizabeth. Known for his great wit and good nature, he enjoyed Elizabethan court life. Harington was also known for the invention of the first toilet with working parts. In his book, *The Metamorphosis of Ajax*, he describes his invention: "This devise of mine requires not a sea full of water, but a cistern, not a whole Thames full, but halfe a ton full, to keep all sweet and savourie." Harington's book offered a diagram along with instructions, cost, and materials needed to build the mighty Ajax, a virtual kit for constructing a toilet.

Although a visionary, Harington's invention was unfortunately not readily accepted by sixteenth-century society. Even

A SIDNEIO SECUNDUS.

His Bodys here by Figure represented,
His Worth is here by Witt to you presented.
Earth was not large inough to hold his Spiritt
Which now y Spatious Heauens doth inheritt .

Sir John Harington (from *The Metamorphosis*
of Ajax, reprinted 1927)

his devoted godmother refused to try his toilet and banished him from court for his poor taste. But his invention caused enough of a stir in society to warrant a reference in one of Shakespeare's plays.

> You will be scrap'd out of the painted
> cloth for this: your Lion that holds
> his Pollax sitting on a close stoole,
> will be given to Ajax.

> *Love's Labour Lost*

Meanwhile, *The Metamorphosis of Ajax* was denied a license for printing. The subject matter may have offended many, but the book proved to be a success even without an official printing license. It went through publication three times. Filled with stories and anecdotes only decipherable to an Elizabethan, it humorously and delicately attacked the essence of the privy. Using poems, Harington educated his readers on privy usage. The following poem gives advice to the Christian pot user in warding off the devil.

> A godly father, sitting on a draught
> To do as need and nature hath us taught,
> Mumbled (as was his manner) certain prayers,
> And unto him the devil straight repairs,
> And boldly to revile him he begins,
> Alleging that such prayers were deadly sins
> And that he shewed he was devoid of grace
> To speak to God from so unmet a place.
>
> The reverent man, thought at the first dismayed,
> Yet strong in faith, to Satan thus he said:
>   Though damned spirit, wicked, false and lying,
> Despairing thine own good, and ours envying,
> Each take his due, and me thou canst not hurt,
> To God my prayer I meant, to thee the dirt.
> Pure prayer ascends to Him that high doth sit,
> Down falls the filth, for friends of hell more fit.

Harington does hit upon a common Christian fear that God is everywhere, even in the toilet. Be careful of prayers uttered while on the pot. An uninvited guest may appear offering untold riches, asking for your soul as collateral.

Harington eventually was restored to his position at court and was able to persuade Queen Elizabeth to install one of his toilets in her palace in Richmond. Harington supplied her with

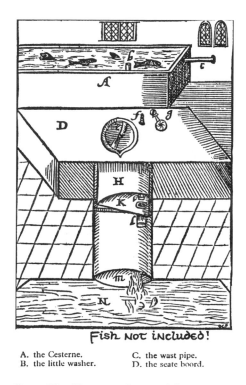

Fish not included!

A. the Cesterne.   C. the wast pipe.
B. the little washer.   D. the seate boord.

The Ajax (from *The Metamorphosis of Ajax*, reprinted 1927)

the first bathroom book in the form of *The Metamorphosis of Ajax* hanging on a chain next to the toilet.

The name Ajax is a mystery. It may be a play on the word *jake's,* commonly used to mean chamber pot. Sir John Harington invented a supernatural character to explain the origin of Ajax. The character named Captain Ajax resembled Ulysses in his powers and pursuits. Anyone who dared to insult the Ajax would be struck down with diarrhea.

-I-

Seventeenth-century England was marked by the plague of 1666 and the Great Fire of 1667. In his journal depicting those awful years, Samuel Pepys spoke of dodging dung in London as it fell from windows above. Jonathan Swift, who was born during this chaotic time, described the streets of London as follows:

> Sweepings from butchers' stalls, dung, guts and blood,
> Drown'd puppies, stinking sprats, all drench'd in mud,
> Dead cats and turnip-tops come tumbling down the flood.

The Great Fire of 1667 proved to be a godsend for the city of London. Many of the squalid neighborhoods harboring disease-carrying pests were destroyed. With a major portion of the city demolished, planners had the opportunity to rebuild according to sanitary codes. The Parliament passed a regulation restricting the height of houses, width of streets, and called for the construction of sewers. However, few people followed Parliament's edicts. In fact, no one seemed to pay attention to government regulations concerning sanitation until the Victorian period.

Of Kingly Deeds

The triumph of monarchy over religion is best exemplified by England's King Henry VIII's break with the Pope. A desire to prove divine right to rule led many of Europe's royalty to outlandish lifestyles. Simple acts of everyday life became elaborate ceremonies. "Going to the bathroom" was turned into a spectacle by rulers such as Louis XIV. Other monarchs merely settled for elaborate chamber pots or the preferred close-stool.

The close-stool, popular during the seventeenth century, was a chamber pot inside a wooden box with a lid. The user simply lifted the lid and sat on top of the pot in the box. The close-stool offered much to the imaginative decorator hiding under furnishings of various kinds. One resembled a stack of books entitled "Journey to the Low Countries." Another stool disguised as a stack of books was titled "Mysteries de Paris." Rather than hide the offending piece, some preferred drawing attention to it. The makers of close-stools employed the use of gold or silver with engravings of birds, landscapes, and Chinese motifs. Velvet, crimson damask, or leather cushioned the privileged rear. Decorations changed with the mood of the day. A death would find the close-stool dressed in black for mourning. James I of England owned a close-stool disguised as an ornamental chest.

Despite the fact that the seventeenth-century Versailles palace contained nearly 274 close-stools, many of the lords and ladies living at court brought their own pots. The practice

Close-stool used by James II of England, Knole House
in Kent, England (Lucinda Lambton/Arcaid)

of greeting visitors while sitting on a close-stool was made famous by King Louis XIV. Few of the foreign ambassadors to the court appreciated the privilege of watching Louis use the pot while trying to conduct business. Courtier Montaigne complained of Louis XIV tending to affairs of state while on the stool "as if it were a throne." An examination after Louis XIV's death revealed he had a huge stomach and the length of his bowels were twice the length of those of a normal man. Possibly the time spent on his "throne" can be explained by his unusual anatomy.

The rulers of Europe had much to say about the hygiene habits of their people. Edicts were issued to dissuade behavior that added to the already nasty atmosphere of the court. Yet, even the royals were known to break their own rules on occasion.

Louis XIV of France on close-stool

Souvenirs of court life have even included accounts of nobles attending to nature in improper but convenient locations.

* Anne of Austria, mother of the Sun King, was caught pissing behind a tapestry in the French palace.
* Henri IV found the filthy condition of the Louvre Palace in 1606 so disgusting that he forbade his nobles to use the corners of the palace to relieve themselves. The punishment for peeing in the palace was a modest fine of a quarter of a crown.
* The poet Berthod found one aristocrat's response to the royal edict humorous. Describing the nobleman's actions, he wrote: "You don't see someone piss; Against a pilar, ha, by my faith; Straight ahead under the statue of the King."
* On August 8, 1606, the dauphin of France issued yet another command on the use of toilets. This time the edict forbade anyone from defecating or urinating in St. Germain Palace. Of course using chamber pots was acceptable but not on the floors, in corners, or in stairways. No one obeyed, including the dauphin. He was caught pissing against the wall of his own bedroom on the very same day he issued the edict.
* The Sun King, Louis XIV, had a friend, the count of Guiche, who was a true opportunist. During a court ball in 1658, the count quietly peed in the muff of his dance partner. Imagine her surprise when her hands slipped into the muff.

-I-

The Duchess Charlotte Elizabeth was part of Louis XIV's entourage. She followed him from his palace in Versailles to his palace in Fountainbleau. The duchess lamented the lack of sanitary facilities in Fountainbleau. Finding only open spaces available for defecating, Charlotte Elizabeth admired the abbess who could pick the time when she had to "go." Only during dark could a person find privacy. The duchess also commented despairingly on the "souvenirs" left behind on the streets by the Swiss guards sent to protect the king.

-ɪ-

During Louis XIV's lifetime, Versailles shone among European courts as the epitome of pageantry. Every aspect of Louis's existence at Versailles was turned into a ceremony. Lords vied for the honor of waiting upon the king as he rose from bed. Dinner was an unrivaled event. The royal repast included four different soups, a whole bird, truffles, and salad—all consumed by Louis in the presence of the entire court.

The end of the day was marked by the *"grand coucher"* and the *"petit coucher"* ceremonies. The *grand coucher,* or the great retiring, involved the ritual of Louis disrobing for bed. Several lords were in attendance hoping to be honored by being allowed to hold the king's candlestick while he undressed. After preparing for bed, all but a few select lords left the presence of the king. These special individuals, having paid up to fifteen thousand louis d'or for the honor, watched as the king began the next ritual, the petit coucher. His Royal Highness bared his royal arse and sat upon the *chaise percée*, close-stool, for the final royal movement of the day.

Lest you think Louis was a flagrant exhibitionist, the seemingly ridiculous ceremonies of Versailles did serve a purpose. Louis XIV controlled his aristocracy by requiring that all the lords spend a large amount of their time living at the court under the king's watchful eye. The rituals reminded them of the king's divinity while providing activities for the lords and ladies, filling the long days and nights. For the aristocrats attending to the king during these private moments, they gained the ear of the powerful ruler. They were able to petition Louis for grants of money or offices.

-ɪ-

Louis XIV's reign is famous for its indulgence of man's natural needs. Sex, food, and hygiene played prominent roles at court. Attending to his own needs Louis rarely was concerned with those of his courtiers. He was fond of traveling from palace to palace in the company of a woman. Without motorized vehicles, these trips were long and uncomfortable. Although many coaches possessed chamber pots hidden under the seats, some of the women chose to wait for a more private opportunity. Louis hated to stop the coach for any reason, purposefully adding to the discomfort of those poor females. A duchess traveling with him to the palace of Fontainebleau nearly succumbed several times to fainting after holding urine in for six hours.

-I-

While engaged in yet another war with Britain and its allies, Louis discovered an able military general in the duc de Vendôme. Louis Joseph, the third duc de Vendôme, was the offspring of Henri IV's illegitimate son by Gabrielle d'Estrees. An egotistical, overbearing individual, de Vendôme presumed upon his royal ties. His feelings of entitlement extended to his flagrant rejection of accepted morals and personal hygiene.

The duc shamelessly required the services of young, attractive soldiers in his tent. His bed was home to all kinds of animals. Dogs slept, wet, and gave birth in the duc's bed. De Vendôme shrugged off accusations of being a slob. "Everyone lives like a pig," he claimed.

When de Vendôme left his bedroom he could be found sitting on his portable close-stool in the military field. While sitting on the pot, he would write letters, eat breakfast, or give orders. He deeply offended the bishop of Parma by receiving the man of cloth as he relieved himself. To make matters worse, de Vendôme stood and wiped himself before the eyes of the appalled bishop.

-I-

Henry VIII is famous for his numerous wives and fondness for a good decapitation. When Henry's relationship with Sir Thomas More turned contentious, the king accused More of treason and sentenced him to death. While waiting for his beheading, Sir (soon to be Saint) Thomas More asked his jailers for a urinal. After peeing in the glass urinal, More examined his urine as if it were a crystal ball. The urine told him what he already knew. He would not live without the intervention of King Henry. Sir Thomas lost his head that day as predicted by his urine.

-I-

Charles II, king of England during the plague of 1666, left London for Oxford in an effort to avoid illness. In his diary, Oxford resident Anthony Wood described the appalling behavior of the monarch and his entourage.

Though they were neat and gay in their apparell, yet they were very nasty and beastly, leaving at their departure their excrements in every corner, in chimneys, studies, colehouses, cellars. Rude, rough, whoremongers; vaine, empty, careless.

Cleanliness & Godliness

-I-

Peter the Great tried to instill in his subjects a Western-style demeanor. He demanded all men shave their beards, on pain of death, in order that they might appear more sophisticated. Russians complied reluctantly. They viewed their beards as symbols of their faith. To please their secular lord, the men shaved their faces. To please their holy lord, they carried their beards in their pockets.

The behaviors surrounding bodily elimination at the Russian court revealed a modest society in comparison to Europeans. The Russian elite considered it impolite to leave the room abruptly to use the toilet. It was also considered taboo to speak of "going to the bathroom." A convenient and immediate excuse was needed to prevent standing around with a full bladder, such as urgent domestic business needing attention.

Peter the Great continued his quest for a sanitized mass society by appointing sanitary inspectors in Moscow. Each inspector was responsible for ten houses. With the passing of Peter the Great, Russia reverted to many of its old ways. Sanitation remained as it was after Peter's death in many of Russia's rural areas. Toilets in those areas consisted of a simple hole in the ground.

--I--

The dangers of bowing in front of royalty are outlined in an incident involving Queen Elizabeth I of England (1558–1603) and one of her courtiers. The earl of Oxford performed the traditional low bow in the presence of the queen. While bowing he let out a foul wind. Horrified at the incident, the earl of Oxford left the country to travel for seven years in hopes of having his faux pas forgotten. On returning to England he was again presented to the queen. Her response, "My lord, I had forgotten the fart."

--I--

Mankind's fascination with the fart dates back to premodern Europe and possibly even earlier. In 1645 a writer in *Wit and Drollery* mused, "Applause is but a fart, the crude; Blast of the fickle multitude." Shakespeare played to a tough crowd.

--I--

Once when King Francis I of France came by to visit his mistress, her lover hid himself in the fireplace. After the king had made love to his mistress, he relieved himself in the fireplace, a common practice. The lover, though urine soaked, escaped.

Pissing Protocol

If spitting chance to moove thee so
Thou canst it not forbeare,

Remember do it modestly,
Consider who is there.
If filthiness or ordure thou
Upon the floore doe cast,
Tread out and cleanse it with thy foot,
Let that be done with haste.

Booke of Demeanour (1619)

-I-

Even in the stench-ridden, filth-laden, and disease-infested cites of early modern Europe, some unhygienic practices were acceptable. It was permissible for horse-wagon drivers to piss on the wheels of their carts. As mentioned before, receiving guests while sitting on the pot was also considered acceptable. However, by the end of the seventeenth century both practices were viewed as inappropriate.

During the reign of John IV and Alfonzo VI, the island of Madeira off the coast of Portugal was particularly strict in matters of sanitation. Partyers caught relieving themselves out-of-doors could be arrested or labeled indecent. Therefore, they peed on porches and in doorways.

Superstitions and Advice

Country residents in seventeenth-century England looking for a place to relieve themselves were told to squat outdoors at a distance of at least a "bow's shot away" from the house. Following the advice diminished the occurrence of human waste seeping into the cottage's well water located near the building.

-I-

Giving advice on bodily functions in his book, *The Breviarie of Healthe for All Manner of Sicknesses and Diseases*, sixteenth-century English physician Andrew Boorde warned readers to avoid the herb thyme. "Tyme causeth a man to make water."

-ı-

The concerns of lovers in seventeenth-century Germany sent many to the village midwife for advice on a myriad of problems. One common solution to the worry of potential sexual problems on the wedding night was to piss through the wedding ring. Obviously if a male could hit the small space in a wedding ring, he had talent. If a woman wants to end without her having to endure a direct confrontation, she might surreptitiously place a tiny sample of her dung in her unwanted lover's shoe. The smell of her, now subconsciously associated with this bad odor, was supposed to drive him away.

-ı-

Doctors commonly examined urine to diagnose illness. A sixteenth-century Italian doctor claimed he could tell much about a person by looking at his pee. The grand duke of Florence consulted the physician to determine his betrothed's state of virginity. The French fiancée agreed to pee in a crystal chamber pot. The physician examined the urine. He declared the French woman a virgin, saying her urine had the same consistency as when she was born.

Early America

The Americas were named for the Italian explorer Amerigo Vespucci, who traveled the Caribbean Sea in 1504. On his first voyage, Vespucci landed on one of the many islands and described the inhabitants in a letter to Soderini, the magistrate of Florence. For Vespucci, the cross-cultural experience fostered a curiosity about the hygiene habits of the natives.

When, begging your pardon, they evacuate the bowels, they do everything to avoid being seen; and just as in this they are clean and modest; the more duty and shameful are they in making water (both men and women). Because, even while talking to us, they let fly such filth, without turning around or showing shame, that in this they have no modesty.

To avoid disease in their villages, Vespucci reported that the natives moved every eight to ten years, leaving the filth behind them.

In 1620, the Puritans traveled to the New World in search of a land where they could establish their own theocracy. Interpreting and practicing the Bible in its most "pure" terms, the Puritans earned a reputation for being religious fanatics. In reality, they were a lively group, and their furniture revealed an appreciation for the finer things in life. Tables were designed for use during the "toilette," the daily act of personal cleansing and beautifying. However, their modesty was revealed in the use of the chamber table. The chamber table hid from sight the pot, which symbolized human needs.

The early settlers to America faced dire circumstances. Many had been urban dwellers and knew little about survival in the wilderness. The newcomers were subjected to extreme weather conditions, such as drought, hurricanes, and harsh winters, as well as attacks by Indians. Certainly one of the more ghastly deprivations faced by early Americans was the absence of the chamber pot. Housing amounted to little more than one-room shacks, and furnishings were scarce. Chamber pots were not in common use until the end of the seventeenth century. The settlers were forced to use the privy in the great outdoors despite snow, sleet, or rain.

As farms began to spring up around the new colonies, so too did privies. Privies were little more than cesspits or holes in the ground. Built downhill from the well, the privies were cov-

Chamber pot in seventeenth-century American home, as displayed at the Museum of American Frontier Culture in Staunton, Virginia (Julie Horan)

ered over as they filled to capacity. A fruit tree was placed over the spot of the privy as a marker.

A few early Americans became rich from trade with England. They were able to afford luxury goods imported from the homeland. Lucky enough to have decent furniture, china, and chamber pots, these settlers turned America into a world of English civility. Houses built in the colonies mimicked the homes the settlers had left behind. Furniture built in the new land copied European furniture. The earliest surviving piece of dated furniture built in the colonies is a chair from Dedham, Massachusettes. Constructed in 1652, the chair contains an enclosed bottom half capable of sporting a chamber pot. The user did not have to leave his seat by the fire to attend to nature's calling. Only the well-to-do could afford this first "lazy boy" chair.

⚛ 4 ⚛

The Age of Enlightenment:
Common Commodes
1700–1800

The eighteenth century produced some of the greatest thinkers in world history. Jefferson, Rousseau, and Voltaire helped to shape modern beliefs on political and social freedom. Unfortunately, these men offered little to advance the disposal of human waste. Thus, the eighteenth century also produced more excrement with no place to go. Chamber pots, close-stools, and privies continued to be the disposal apparatus of choice. In the ditches and streets of London accumulated the carcasses of dead animals left by butchers, human waste thrown out windows, and the daily garbage produced by every household in the vicinity. As waste was cleaned from the streets, more replaced it. Sanitation workers carted the refuse outside the city limits to greet visitors to London.

The golden age of our forefathers contained some minor improvements to daily living conditions. The 1700s was a period of transition before the Victorian era began. Privacy in attending to nature's needs gained a permanent place in the

expectations of eighteenth-century Western society. Privy closets housing chamber pots or close-stools replaced Louis XIV's
practice of holding court while sitting on the pot. Famous furniture manufactures designed pieces to hide the presence of
the chamber pot. Most exciting of all, more houses were being
connected to the main sewer drain serving London. Primitive
water closets (indoor privies connected to sewers) were
installed in a few of the wealthiest homes. Finally, by the end
of the century the first "modern" water closet with a working
valve was invented—the precursor to the nineteenth-century
toilet. And of course, we cannot forget the invention that has
become a prominent symbol of French society: the bidet.

Pots

> Presumptuous pisse-pot, how did'st thou offend?
> Compelling females on their hams to bend?
> To kings and queens we humbly bend the knee,
> But queens themselves are forced to stoop to thee.
>
> *Scatologic Rites*

Improvements in the process of making china porcelain
turned the dull chamber-pot industry into one turning out
works of art. The miracle came from the use of silica in producing a durable white porcelain that could be painted and
printed upon. Chamber pots became so beautiful, it was a
shame to squat over them. Some of the designs incorporated
humorous commentary.

-꘏-

A few months after signing the Declaration of Independence,
Benjamin Franklin sailed for France to convince the French
people to become an ally of the new nation in the
Revolutionary War against Britain. Franklin assumed the role of

Design on chamber pot belonging to the Duke of Wellington
at Stratfield Saye, England (Lucinda Lambton/Arcaid)

a celebrity in Paris. People crowded the streets to see the
witty, eccentric American. The visiting statesman agreed to sit
for several portraits while on his mission. But the portrait the
jovial Franklin would have appreciated, without taking offense,
was his caricature decorating a chamber pot. The inscription
on the pot read *"Eripuit coelo fulmen, sceptrumque tyrannis."*
Roughly translated to mean "he who takes away with a thunderbolt the authority of tyranny." It was a dubious honor considering Franklin's face was being pissed on by the French pop-
ulous. Customary to his easygoing manner, Franklin reportedly
just smiled when he saw his mug on the chamber pot.

A playfully jealous King Louis XVI discovered that one of his
lovers had a crush on Benjamin Franklin. As a birthday present,
the king sent the lady a chamber pot with Franklin's mug on it.

The chamber pot continued to be a nuisance to some and an offense to others. The French painter Jean-Baptiste Greuze (1725–1805) was hardly pleased with the chamber pot when his wife decided to use it as a weapon. Battered about the head with the pot, Greuze sustained only minor injuries. Although a handy weapon for irate housewives, the following anecdote reveals why Jonathan Swift considered the chamber pot an affront to the sensibilities of refined persons.

The inconvenience of leaving the house to use a detached privy in the backyard caused many people to rely on the pot. In his book *Direction to Servants*, published in 1745, Jonathan Swift complained of what he viewed as a repugnant practice: dependence on the chamber pot. He offered advice to servants on breaking the offensive habit of their masters.

> I am very much offended with those ladies, who are so proud and lazy, that they will not be at the pains of stepping into the garden to pluck a rose, but keep an odious implement, sometimes in the bed chamber itself, or at least in a dark closet adjoining, which they make use of to ease their worst necessities; and you are the usual carriers away of the pan, which maketh not only the chamber, but even their clothes offensive, to all who come near. Now, to cure them of the odious practice, let me advise you, on whom this office lieth, to convey away this utensil, that you will do it openly, down the great stairs, and in the presence of the footmen: and, if anybody knocketh, to open the street door, while you have the vessel in your hands: this, if anything can, will make your lady take the pains of evacuating her person in the proper place, rather than expose her filthiness to all the men servants in the house.

His advice to newlyweds: "Keep them to wholesome food confin'd, Nor let them taste what causes wind: 'Tis this the sage of Samos means, Forbidding his disciples beans."

-I-

To accommodate their guests at parties, hosts rented chamber pots for a nominal fee. As the guests became drunk, pots were broken under the clumsy weight of the users. The sign of a successful party was the number of broken chamber pots. The Irish and Scottish were famous for the number of pots they had to replace after a party.

-I-

The end of the eighteenth century marked a change in attitudes toward bodily functions. While bodily functions had been viewed as natural and inevitable, they were now considered something to be hidden and ignored. Furniture made by the famous Hepplewhite and Sheraton served as hiding places for the indecent pot. Night tables, found primarily in the bedroom, consisted of a cabinet space able to hold one or two chamber pots. Variations on the night table held the basin and ewer for morning wash. Made from fine wood and carved in the fashionable style of the period, the tables did their job by looking too handsome to be hiding someone's excrement. Unfortunately, nothing could mask the sickening smell.

Close-Stools

Aristocrats were reluctant to give up their close-stools for the more sanitary privy. The close-stool represented an individual's throne, decorated to suit the extravagant tastes of its owner. Louis XV and his successor Louis XVI took pride in their close-stools. Referred to alternately as the "chair of affairs" or "necessary chair," this piece of furniture seemed to honor the bowel movements of its user with ostentatious designs. Louis XV (1710–1774) owned a close-stool made of black lacquer and painted with Japanese landscapes and birds in gold. It contained inlaid mother-of-pearl and bronze fittings.

The inside of the box was made of red lacquer. The seat invited long periods of meditation with its gentle padding of green velour.

The two mistresses of Louis XV kept him busier than the administrations of state. Richly benefiting from their lover, both women owned elaborate close-stools. Madame de Pompadour (1721–1764) was so delighted with her stool, she bestowed on its maker a hefty annual pension in gratitude. Her successor, Madame du Barry, owned a stool decorated with blue motifs, red stars, and black lines inlaid on a white background. Gold-embroidered blue velvet trimmed the stool. The arms and feet of the chair were made of gold, and the seat was covered with Moroccan leather. The pot under the seat was made of silver.

-I-

Modesty caused the close-stool to move into the closet for privacy. The once unisex device developed into yet another agent for separating the sexes. At a party in Paris in 1739, close-stools were placed in small cabinets containing the inscriptions Ladies and Gentlemen. Naturally, the line for the ladies' cabinet was twice as long as the one for the gentlemen's.

Provincial Privies

Indoor/outdoor privies maintained their status as one of the big three of elimination devices (privies, chamber pots, and close-stools). Without running water, little could be done to improve the privy. Ventilation of the space posed a big problem for house owners.

A visit to the private home of Thomas Jefferson is further proof of the man's genius. A keen architect and inventor, Jefferson built into his house automatic doors, a clock that tells the days of the week, and skylights. A compliment to Jefferson's well-known imagination is the apocryphal story that Jefferson built a tunnel to haul the privy waste from the house to a sewage pit on wheeled carts. Tunnels, which measured 160 feet in length and connected two privies on the first floor and one privy on the second, did exist at Monticello. However, no tracks or wheeled carts were unearthed. The extensive tunnels assisted in ventilating the privies by creating a draft of air that escaped either through a skylight or a connection with the chimney flue.

Besides the three privies in the house, Monticello retained two outhouses, or "necessaries," as Jefferson called them. Responsible for emptying the contents of the house privies was a slave, whom Jefferson paid for performing the unwelcome task.

-I-

In Europe and America during the eighteenth century and well into the nineteenth century, most privies continued to be found outside, usually near the garden. Some were so decorative they could be mistaken for small houses.

Ranging in appearance from boarded shacks to marble temples, privies reflected the wealth of their owners. Generally built a short distance from the house, sometimes hidden in the garden, it was possible for the user to claim she was going to "pluck a rose." Apparently, a common belief held that the family that defecated together, stayed together. During this period, privies generally accommodated several people at a time. Some privies even included a child-sized hole.

-I-

Travelers in Europe and America during the eighteenth century found food and a place to sleep in taverns along the highways. With the locals, the travelers could enjoy a homecooked meal and alcoholic beverages. Drunk guests were not uncommon in taverns. John Michie, the proprietor of Michie Tavern

Six-seater privy at Chilthorne Dormer Manor, Somerset, England,
mid-1800s (Lucinda Lambton/Arcaid)

in Virginia in 1784, complained of the necessity of rescuing drunk visitors from the outhouse after they had fallen into the privy hole. In addition to a box holding dry corn cobs for wiping, Michie hung a rope from the ceiling of the four-seater outhouse to allow anyone falling into the hole to climb out. He posted a notice on the "necessary house" that read: "Notice Ye All, If ye bottom falls through ye seat, Do not call the proprietor, Use ye rope to pull ye out."

Bodacious Bidets

Although not a means of disposing of human waste, the bidet, first appearing in France during the eighteenth century, deserves mention. The origin of the bidet remains a mystery. The etymology of the word *bidet* dates back to 1534, referring to a donkey or horse. In the 1700s, the bidet came to be associated with the device used to clean one's bottom after defecation. The choice of bidet as a name for the device came possibly from the action of straddling the elongated bowl as though it were a donkey or a horse.

The bidet quickly developed a reputation as extravagant and sexual. In 1751, the genital-cleaning device was alluded to as the *"indulegence of Madame de Pompadour."* Its reputation became notorious with the publication of Thomas Smollet's travel letters from 1763. Smollet wrote, "Will custom exempt from the imputation of gross indecency a French lady, who shifts her frowsy smock in presence of a male visitant, and talks to him of her lavement, her medicine, and her bidet!" To the English and many other Westerners, the French bidet was considered a product of French immorality and used for cleansing genitals after sex.

Water Closets: Looking Toward the Future

The first water closets were little more than privies moved indoors and built into a corner niche or small closet. To flush, one pulled on a handle to open a trap; water then escaped from the cistern above and washed the contents down. Requiring a connection to a sewer, an expensive enterprise, and a steady supply of water, the water closet appeared in only a small number of wealthy homes.

Queen Anne of England (1665–1714) possessed a water closet built into a small space off her dressing room in Windsor Castle. The seat of the privy was made of marble. Although water washed the waste down the privy, the absence of a trap and valve meant it was rarely cleaned adequately by the water, and odors escaped constantly from the sewer below.

The emergence of the water closet in houses was greeted with suspicion by many in the eighteenth century. The writer Horace Walpole believed a person owning a water closet was morally depraved. Describing a visit he made to the house of Aelia Laelia Chudley in 1760, he said, "But of all curiosities, are the conveniences in every bed chamber: great mahogany projections ... with the holes, with brass handles, and cocks, et cetera—I could not help saying, it was the *loosest* family I ever saw!"

The late 1700s marked the beginning of "true" modern sanitation. Improvements in the simple water closet led the way to the modern toilet of the next century. Beginning with an invention by Alexander Cummings and culminating in adjustments by Joseph Bramah, the late seventeenth century offered the promise of a water closet without odor.

In 1775 Alexander Cummings, a London watchmaker, received the first patent for a water closet. His design greatly improved upon the fledgling invention by Sir John Harington two hundred years earlier. Cummings's design incorporated Harington's use of gravity to aid the flow of water. But more importantly, Cummings had the brilliant idea of using a valve trap to secure the area between the bowl and the outtake pipes. The closed system meant less smell.

As Horace Walpole's reaction illustrates, the new water closet was not an instant success. Besides the fear of decadence, few people were ready to part with the tried-and-true tradition of the close-stool and chamber pot.

-ɪ-

Not to be confused with the holy elite of India or the cow, Joseph Bramah constitutes an important historical figure for his contributions to the toilet. Thanks to Bramah we have the swirling action of water in the bowl, which helps clean out the contents, and an improved flap system. The biggest problem with Cummings's water closet was his design of the flap, which slid open when a mechanical arm was pulled. Unfortunately, the few water closets in use were installed in privies or outhouses separate from the main house. Cold weather wreaked havoc on the slide flap. You can imagine the buildup when the flap was frozen shut.

Bramah's hinged-flap design of 1778 allowed the contents to empty and then sealed the space afterward. By 1797 Bramah had produced six thousand water closets. Not a large number considering the population of England was about eight million at the time. Distribution was slowed by the fact that London contained no standard sewer system. If a neighborhood was lucky enough to have a sewer, it was probably different in design from the next neighborhood. Joseph Bramah went on to invent the hydraulic press and the lock. His water closet design continued to outperform challengers for the next ninety-eight years.

Smelly Moments in the 1700s

Every period of history offers the student colorful stories about sanitation and hygiene practices. The eighteenth century was no different.

--I--

Mozart wrote some curious love letters to his cousin. One included the sign-off "Well, I wish you good night but first shit into your bed and make it burst."

--I--

The first johnny on the job was a human operation. Traveling the streets of Edinburgh, Scotland, plying his trade, Johnny offered his customers "privacy" while relieving themselves. He wore a large black cape and carried a chamber pot, crying "Wha wants me for a bawbee?" A customer approached the John, paying him a halfpenny. While the customer squatted over the pot, Johnny covered him with the large cape.

--I--

A caricature published in 1796 called "National Conveniences" revealed English pride in their water closet and their contempt for other Europeans. The cartoon depicted the

English with their water closet, the Scottish with a bucket, the French with a latrine, and the Dutch using a lake.

-ː-

The ocean provided the most convenient cesspit available to mankind. The challenge for sailors lay in devising a way to use the ocean safely. Smaller ships relied on buckets that were dipped into the water after use. Larger ships contained privies that collected the "offerings" in vessels and then dumped them into the ocean. The privies offered privacy, which was highly prized on a crowded ship. Another arrangement found on ships included a seat located on the bow (the head of the vessel) that emptied directly into the water. Sailing ships were pushed from behind by the wind. Placing the privy seat on the bow of the ship, overhanging the water, prevented the droppings from flying back onto the ship. An added feature, the smell was kept to a minimum. The "head" became synonymous with "toilet."

But many dangers befell unsuspecting sailors using the privy seat. It is not known how many men may have lost their lives falling from the head. While calm weather presented no problems, high winds and waves often combined to soak the innocent user. In addition, a sudden upward gust could result in a person being covered with the very waste he was dumping. Some sailors were known to climb onto the seat stark naked in order to protect their clothing.

-ː-

European cures for health ailments in the eighteenth century were little improved from previous centuries. Patients suffering from strokes were told to swallow a glassful of urine collected from a healthy person. Mixed with salt, the concoction was designed to purge the body of bad "humors." It was believed that cataracts would disappear if human dung was dried, powdered, and blown into the eye of the sufferer.

-I-

Servants attended to every need, requested or anticipated, of the nobleman. A footman followed his master so closely that he earned the name "fart catcher." The servant expected to attend to all of the private needs of his master, and he considered himself to possess the best job in the household. As manservant to the master, he assisted his lord with dressing and toiletry. The servant provided a clean towel, warm water, and a basin for the morning ritual of cleansing. In the evening, the "lucky" servant provided a cushion for the board covering the privy, or "house of easement" as it was often called. After making sure the privy was clean and did not smell, his last act for the evening was to place a urinal next to the bed for nightly use by the master.

-I-

Politically, the bourgeoisie, or middle class, used the poor to advance themselves against the aristocracy during the French Revolution. However, the middle class had opinions concerning the poor and the rich that went beyond governing. The cosmetics, outrageous wigs, and heavy perfume worn by aristocrats were thought to be indicative of their lazy, indulgent nature. To the other extreme, the peasantry were viewed as superstitious fools who believed dirt and urine were beneficial to personal hygiene. Bourgeois writings, such as the *Journal de Santé* in the 1780s, called for the cleaning of the environment and promotion of personal health. Unfortunately, the call was not heeded. It was easier to pursue a political than a sanitary revolution.

-I-

Benjamin Franklin was famous for his many interests and pursuits. He invented bifocal lenses, published several books, and conducted science experiments with electricity. Another first for Franklin was his design of a second-story privy in his Philadelphia home. The drains of the privy on the second floor were located directly over the first-floor privy.

❧ **5** ❧

The Victorian Age:
Prim and Proper Piss Pots
1800–1920

"A sewer is a cynic. It tells all."
Victor Hugo, *Les Miserables*

Moral propriety marked the era bearing Queen Victoria's name. Books published in nineteenth-century Britain and America described the proper way to serve a dinner or have sex with your husband. Victorians may have been clean in their souls, but their bodies and city streets were filthy. Slums were home to millions of people who worked in factories for pennies a day. Streets were filled with garbage and human waste from overcrowded urban dwellings. City maps from some western American states showed the location of dung, attesting to its permanence as part of the landscape. Cleanliness was not a priority of the government. A street in London went uncleaned for fifteen years.

Prior to the nineteenth century, sanitation regulations instituted by state governments in the Western world occurred infrequently and were rarely enforced. Laws concerning disposal of excrement generally appeared only after a major epidemic of disease. But the remarkable influx of people to city centers during the Industrial Revolution of the nineteenth century, first in Britain and then in the United States, moved citizens, charities, and politicians to call for government intervention for a healthy environment. Responding to the increasing calls for action, the chancellor of the Exchequer of Britain proclaimed, "Sanitary reform is a humbug." In Lyon, France, an official at the Royal College claimed that loos would lead to the end of respectability and morality.

Not until a series of smallpox, cholera, and typhoid epidemics occurred during the mid-1800s in Europe and North America did people begin to take notice and make sanitation improvements. The slow response in establishing sanitary regulations as a means of curbing disease was due to debate in the scientific world as to whether a connection between poor sanitation and disease existed. In 1854, a British doctor, John Snow, was able to trace an outbreak of cholera to a neighborhood water supply contaminated with feces. However, there was no concrete evidence to support Snow's claim, and it was dismissed. Without verification of a connection between disease and sanitary conditions, British lawmakers refused to infringe on citizens' property by passing a law to build sewers. When possible, such as after a fire, city planners included provisions for sewers in rebuilding. However, only minor changes could be justified until science could prove the connection of poor sanitation facilities to disease. Finally in 1883, scientist Robert Koch confirmed the beliefs of many sanitary reformers by isolating the germ for cholera under the microscope.

By the late 1800s, the golden age of toilets had arrived. Suddenly, toilets were on everyone's mind. Architects earnestly began incorporating flush water-closets (toilets) into building plans. Inventors and entrepreneurs scrambled to corner the new market. However, because of Victorian modesty, dealing with the body's personal needs demanded creative minds.

Squeamish Squalor

Understanding the mammoth task faced by city officials, engineers, and inventors in ridding Victorian cities of their foulness requires an appreciation of the truly squalid conditions in which people lived. Even cities in pristine Australia did not escape problems arising from increasing population and the waste that accompanied it.

Doctors in Melbourne, Australia, during the nineteenth century advocated replacing trenches with a modern sewage system. The city had been dubbed "Smelbourne" because of the odor arising from open sewage gutters running through town. But their pleas fell on deaf ears until the 1890s when the Melbourne Metropolitan Board of Works formed to combat the pollution of nearby rivers by sewage. Despite a population density of only six persons per acre as compared to London's forty persons per acre, Melbourne's mortality rate exceeded that of London. The installation of modern sewers and drains decreased contamination found in the water, effectively lowering Melbourne's mortality rate.

-·I·-

As always, the poor were the last to benefit from sanitary reforms. While sewers and privies were increasingly common in neighborhoods serving the rich and middle class, the poor used common privy pits found in tenement buildings or their backyards. It took 2,300 night-soil carts to empty the tenement privy pits in Paris. In Limoges, there were no winners. Gutters continued to collect raw sewage and garbage as they had in medieval times.

As the population of Paris increased, modernization of the small medieval streets and buildings became necessary. Emperor Napoleon III (1808–1873) supported the redesign of Paris in the mid-1800s to widen streets to enable the passage of a large army. The Opera House and Les Halles, the town market, were built during this period. More importantly, by 1870 most of the city's 805 kilometers of streets had sewers beneath them. The discharge from the sewers flowed into the Seine River, which then transported the foul waste to communities downstream.

―I―

Houses in Manchester, England continued to rely on the "pail system" until the turn of the century. This consisted of petroleum casks placed in front of every house for collecting the inhabitants' waste from chamber pots. Specially colored casks were placed in front of the houses of sick people to warn of disease.

Charles Dickens in Britain and American reporter-photographer Jacob Riis in his book *How the Other Half Lives* (1890) raised people's consciousness with their descriptions of children growing up in the slum tenements of newly industrialized nineteenth-century cities. The images of despair awakened the hearts of middle-class citizens. Reform movements developed calling for changes in the environment and morality of the poor. The Women's Christian Temperance Union (1874), formed in Cleveland, pressured local governments to abolish the sale and use of alcohol, believing it to be the cause of crime, disease, and broken homes. Political reform, through the Progressive Movement of the late eighteenth century in the United States, secured national laws concerning child labor and instigated a push for sanitary reform. The intensity behind nineteenth-century social-reform movements matched the growing intolerance of filth.

Poverty can lead people to desperate acts. In Victorian England the poor scoured the sewage and garbage dumps along the Thames during low tide, searching the muck for food or valuable items. So prevalent was the behavior that they became known as "mudlarks."

Working children were common in Victorian England. Factories were full of them. Eight-year-olds were employed for pennies a day. Some children worked as "pooper-scoopers" around the city. They roamed the streets of London collecting the dung left behind by dogs and sold it to tanners for use in tanning leather.

-I-

Despite the pressures of overcrowded cities, rampant disease, and pure stench, many people managed to just ignore the unpleasantness. Victorians believed in circumspect behavior: Anger or any display of extreme emotion was abhorred. Thus, Victorians tried hard to deny natural emotions and bodily functions. "You won't enjoy sex, dear. Just close your eyes and think of Britain." Such was the stereotypical advice given to daughters of marriageable age. Of course, stereotypes are seldom true. But when it came to relieving themselves, the Victorians went to great lengths to hide the action.

Furniture designs during the nineteenth century hid the presence of the necessary, but offensive, chamber pot. The furniture produced was reminiscent of James Bond's toys. Gadgets and secret doors were hidden in every corner of a seemingly innocuous cupboard. In 1833 John Loudon described the "wash-hand stand" as enclosed in a bureau with the washstand opening to accommodate a basin, soap dish, and a comb box. The underside of the cover could be raised via a rack and horse. Below was a space capable of holding an ewer, a basin, and a chamber pot. George Jennings designed a combination bidet, footbath, sitz bath, and commode pail. The ambitious invention was reversible and fit in one piece, taking up little space.

Close-stools and night tables built in the nineteenth century tried to draw attention away from the act of defecating and toward the enjoyment of music. One type of close-stool played chamber music when its lid was lifted. A night table provided music when its door was opened for access to the chamber pot. As Victorians found bodily functions embarrassing, the musical cover-up relieved the anxiety of being heard while relieving oneself.

-ー-

The Hundred Years War between France and Britain ended in 1453, but the war of words has continued through the centuries. During the Victorian era, the British often accused the French of licentiousness and rudeness. The Frenchman Louis Simond, traveling through Britain in 1810, returned the compliment. Writing of his travels, Simond described his horror at the uncivilized sight of a chamber pot or close-stool occupying a dining room corner in many of the English homes he visited. Simon wrote that the British used the facility with no show of modesty and often while in conversation with guests. Apparently, by that time, the French had exported their custom of "French Courtesy," the practice of receiving guests while sitting on the pot.

Continuing the condemnation of the odious, gauche British, Jean-Anthelme Brillat-Savarin described the British reliance on the chamber pot as follows.

> The 'curious facility' had been enjoyed for a great number of years, not only by diners but also by gentlemen travelling in their coaches (silver or pewter), ship's captains (silver or pewter) and judges on the bench (pewter or occasionally porcelain vases). Nowadays the facility is rarely made use of although still, on an English night, the moon

shining over an estate will illuminate a crocodile of dinner-jacketed figures, wreathed in cigar smoke, picking its way carefully towards the compost heap, or lined up like a firing-squad opposite the rose-bed."

-·I·-

Morality dominated French life in the mid–nineteenth century as it did the lives of Englishmen under the rule of Queen Victoria. The attorney general of France, Pinard, believed his position as a public administrator included defending public chastity. He targeted immorality in all areas of society. As part of his duty, Pinard arrested Gustave Flaubert in 1857 for writing the novel *Madame Bovary*.

Operating on a tip from a competing manufacturer, Pinard investigated a manufacturer of chamber pots for obscenity. The manufacturer had produced a chamber pot with a large eye painted inside the bottom of the pot. I SEE YOU was the inscription that accompanied the eye. Pinard was not amused. The attorney general arrested the manufacturer, who was then sentenced to one month in jail.

English aristocracy taking a break

Pinard had a counterpart in the fight against licentiousness. Perhaps the inspiration for the obscene chamber pot, Madame de Celnart recommended to her readers that they close their eyes when cleaning their private parts.

-I-

The French may have lost their sense of humor, but the British had not. They tweaked their own prudish tendencies in a euphemistic poem that created a stir with its allusions to using the toilet. An anonymous Victorian poet offered a humorous prayer directed to the ancient Roman goddess of the sewer, Cloacina.

> Oh Cloacina Goddess of this place,
> Look on thy servant with a smiling face
> Soft and cohesive let my offering flow
> Not rudely swift nor obstinantly slow.

Temples of Convenience

Our Heros

Astronauts were to be the heros of the twentieth century; but in the nineteenth century, toilet inventors were the giants that walked among men. Even royalty responded to the lure of the heroic plumbing entrepreneurs.

The drive to reform sanitation was encouraged by the Prince of Wales in 1871. While visiting the house of the countess of Londesborough, the prince caught typhoid. The disease was traced to the house's sewer drains. The Prince of Wales nearly succumbed to typhoid as his father Prince Albert had in 1861. Two other persons visiting the countess died from the disease. The legitimacy of sanitation reform was born when the prince supposedly uttered, "I should like to be a plumber if I were not a prince."

-I-

Building on the groundbreaking inventions of Cummings and Bramah, three British entrepreneurs parlayed the concept of a toilet into a viable industry. The Three Musketeers of plumbing were George Jennings, Thomas Crapper, and Thomas Twyford. They proved to be the most successful of toilet traders in England. Rather than designing an original system, these nineteenth-century inventors tinkered with the existing "Bramah" until each problem was solved.

George Jennings designed a siphonic wash-down closet that greatly increased the pressure of the water entering the toilet bowl. As a result, the rush of water emptied and cleaned the bowl better than previous models had. Thomas Crapper designed a pull-chain that worked in conjunction with a valveless cistern, thus decreasing noise and preserving water. Thomas Twyford contributed to the appearance of the toilet. Removing the wooden chair covering the metal working parts of the water closet, Twyford encapsulated this mechanism in porcelain. The move to porcelain provided an aesthetic and functional addition to the toilet. Porcelain bowls made cleaning much easier. In addition, the bowls designed by Twyford were works of art. They were molded into lions, dolphins, and flower motifs.

The achievements of these men were no small matter. The water closet they had inherited consisted of a cast-iron bowl cleansed by a cistern of water released by a handle and emptied directly into the drain below. There was no effective means of preventing sewer gas from escaping into the house. In fact, hotel guests had no need to ask for the location of the "loo" on their floor. They merely followed their noses.

-I-

Construction flaws produced many of the sanitation problems of late nineteenth-century homes. All drains, kitchens, bathtubs, and water closets in the house connected to a main soil pipe that led to the sewer pipe. Unfortunately, the drains had no valve traps. Stench and bacteria permeated the house and water. In addition, poor ventilating systems contributed to the odors.

The smell of backed-up sewer gas was offensive even to the conditioned noses of the nineteenth century. Plumbers tracked breaks in drains by tracing peppermint oil released into the pipes. Where the peppermint smell overwhelmed the normal stench a break was indicated. Faulty drains and sewers were more than a mere nuisance. The buildup of methane gas created a dangerous situation for homeowners and plumbers inspecting obstructions.

Poor design and lack of ventilation in Victorian plumbing made a plumber's job a dangerous one. If a plumber did not die from disease contracted on the job, he was apt to die from explosions. Rats became the friend of the plumber. The appearance of rats in the sewer indicated all was well. Dead rats meant poisonous air and a possible explosion if a match was struck. Primitive water closets provided breeding grounds for water-borne diseases such as cholera. Houses hooked up to the sewer system offered no respite, as the first sewer drains were made of brick and moved slowly if at all.

-I-

Edwin Chadwick was Britain's leading health reformer in the nineteenth century. In 1842, Chadwick published a general report entitled *The Sanitary Conditions of the Labouring Classes in Great Britain*. This work is credited with highlighting the connection between overcrowded conditions, lack of adequate disposal for human waste, and diseases such as dysentery, cholera, and typhus. Most important, Chadwick proposed modern sewers to correct the problem and emphasized that prevention of these diseases could be obtained for as little as 4 pounds sterling per house.

In response to Chadwick's report and an outbreak of cholera in 1848, London's government conducted a survey on the conditions of its sewers. The results were astounding. One sewer line produced dead cats, dogs, and rats, the remains from a slaughterhouse, horse dung, ashes, pans, stoneware, bricks, and other rubbish. Following the report, London passed its first Public Health Act. The act failed to require strong action to combat the city's filth. But successive sanitary acts put in place the regulations required for disease control. The most important regulations called for local governments to make provisions for human-waste disposal through use of sewers and the cessation of water companies selling water drawn from areas on rivers where sewage was dumped. In 1872, the London legislature passed the Metropolis Water Act, which condensed eight separate water companies into one. Moving toward a standardized sewage system was paramount in the effort to serve all inhabitants of the city. However, technical problems threatened to disrupt the common sewer. Water closets with a water-tight valve were required to prevent standing water, the breeding ground of disease, and loss of water through valves that would not refasten. Instead of saving London from epidemic diseases, an inadequately designed sewer system could act as a conduit of disease throughout the environs.

-·I·-

Responding to the call for an improved water closet was the now-famous Thomas Crapper. Crapper perfected the valve system of the toilet with his 'Valveless Water-Waste Preventer' in 1884. As its name implies, the Valveless Water-Waste Preventer prevented the loss of water from cisterns equipped with loose-fitting valves. The novelty of Crapper's design was in the way the water could automatically refill without a slide valve. The pull-chain, attached to a circular chamber above the cistern, unleashed water when engaged. The water traveled up a pipe, displacing the pipe's air. The force created from the water's movement emptied the cistern tank, sending the water into the flushing bowl. The design became known as the Pull and Let Go because the user need not hold the chain until the contents of the tank emptied. The importance of his work in the sanitation field won Thomas Crapper a knighthood from the queen. However, Thomas Crapper is not a household name because of his obscure patent. We have the American army to thank for making the name Crapper infamous.

Stationed in Britain during World War I, American GIs noticed a pattern in the water closets of London. The majority of WCs bore the stamp of Thomas Crapper & Company. The GIs returning from England after World War I brought with them a new word, "crap" or "crapper." The ignoble origin of the word derives from young men saying, "I'm going to the Crapper."

Before Thomas Crapper set up shop in the Marlbourgh neighborhood of London, George Jennings convinced Prince Albert to allow him to display a water closet in the Crystal Palace Exhibition of 1851. Jennings believed that "the civilization of a people can be measured by their domestic and sanitary appliances." Unfortunately, the Victorians were slow to accept the exhibition of WCs in public. When a replica of the Crystal Palace was built, Jennings's WCs were not installed.

But George Jennings saw an opportunity to advance public hygiene by connecting his WCs to the new sewage system built in 1859. He persisted, and by the turn of the century public toilets bearing the "washdown" closet invented by Jennings graced several railway stops, parks, and other public thoroughfares around the world in places such as America, Argentina, South Africa, and Mexico. As the first modern public toilets, they were disguised behind iron arches, pillars, panels, and lamps. Some public toilets were built underground to avoid offending "delicate minds." Jennings's vision of the public toilet included an attendant who, for a fee of a penny, would provide a clean towel, comb, and clean each seat after use with a damp piece of leather. In addition, a shoe shiner would be on the premises for the patrons' needs.

Travelers to Europe will see the remains of public urinals built at the turn of the century. Located at the intersection of busy streets or in a park, the urinals, commonly called pissoirs, accomodated men only. With no doors and merely a drain in the floor, the pissoirs enabled men to quickly attend to business and be on their way.

-I-

The favorite toilet of royalty in the nineteenth century was the Optimus. It was installed in Buckingham Palace, Windsor Castle, and Hampton Court Palace. In addition, the czar of Russia, the king of Thailand, and the duke of Wellington all owned an Optimus. Designed by Stevens Hellyer in 1870, the Optimus was a valve closet that concealed the pipes by placing the whole apparatus in a chair.

-I-

The trials and errors in producing a WC may surprise the modern reader, who takes this device for granted. Constant testing of new inventions improved the quality of the toilet. To determine a toilet's ability to flush matter effectively, objects of different sizes and weights were flushed down it. Jennings tested his Pedestal Vase by flushing ten apples, one flat sponge, and four pieces of paper. It was a success. The Pedestal Vase won the gold medal at the Health Exhibition of 1884. An entrepreneur named Shanks developed a cheaper version of Jennings's invention. A cheaper test was needed. He grabbed the hat off his assistant's head, flushed it down the toilet, and cried, "It works!"

Jennings deserves to be known as the father of the toilet. His "closet of the century" improved on the siphonic system of flushing and became the model for modern toilets. The siphonic action allowed for a fast flush followed by a slow flush emptying the bowl more efficiently.

-I-

Besides the problems with trapped sewer gas, exploding sewer lines, and disease, the most embarrassing dilemma associated with the new water closets was the noise they created. The loud noise emitted upon flushing the WC announced to the entire neighborhood that someone had used the toilet. Statements such as "Excuse me while I take the scones out of the oven" fooled no one when the house vibrated from the impact of water gushing through the pipes. Discovering a way to silence the toilet became paramount to the Victorian public. Unfortunately, success eluded these modest citizens. A kindlier, gentler, quieter toilet was not invented until the next century.

Problems associated with the water closet—lack of a sufficient water supply, smell, and noise—led to the development of alternative methods. The Reverend Mouler of Britain invented the earth closet. Using the "silent" earth, or ash, rather than noisy water to clean the bowl, the earth closet appeared very similar in design to the water closet. A pull-handle opened a hopper releasing earth to cover the excrement lying in the bowl beneath the seat.

-I-

Dr. Vivian Poore invented an ecological closet that he described in his book *The Dwelling House.* Similar to the earth closet, the ecological closet used earth to catch the dung. In addition, a box was provided for composting the human dung into fertilizer. To complete the operation, Dr. Poore suggested growing a clatlava tree nearby in order to use its leaves for toilet paper.

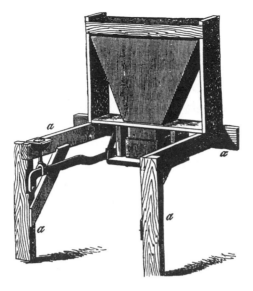

Reverend Moule's Earth Closet patent, 1860

French Sewers

Early sewers in France resembled little more than street gutters. Because they were open and relatively shallow, the sewers overflowed during heavy rains. Men called *pontonniers volants* took advantage of the situation to earn some money. They placed planks over the sewers for pedestrians to cross the street, for a small fee. These early sewers numbered few and often were private. As a consequence, the French sewers were rarely maintained.

As the city of Paris began to build sewers beneath city streets, the repulsive tunnels became sanctuary and home to the city's criminals. After mugging pedestrians or stealing from shopkeepers, thieves would slip into the depths of the sewer confident that no one would follow them. Citizens feared the sewers, viewing them as a criminal underworld. Victor Hugo found the atmosphere of the sewers synonymous with a civilization. In his book, *Les Miserables*, Hugo wrote, "The history of man is reflected in the history of sewers. . . . The sewer of Paris had been a formidable old thing. It had been a sepulchre; it has been an asylum. Crime, intelligence, social protest, freedom of conscience thought, theft, all that human laws prosecuted or have prosecuted, was hidden in this pit. . . ."

The sewer's reputation for subterfuge expanded during the French Revolution. The revolutionary, Jean-Paul Marat (1743–1793) was rumored to have hid during the Revolution in the sewers of Paris. The theory of his hiding place fit nicely with his diseased appearance of open, oozing sores and severe body odor.

In *Les Miserables*, Victor Hugo described the Parisian sewers of the early eighteenth century through the character of a city official. "The lanterns would scarce burn in the mephitic atmosphere. From time to time a sewerman was carried away unconscious. At certain spots there was a precipice; the soil had given way, the tiling had crumbled, the drain had become a cesspool; nothing solid could be found."

Summing up the environment of Paris during the nineteenth century, Edwin Chadwick, a British public-health leader, stated, "Fair above, foul below." Although Chadwick was undoubtedly correct in his assessment, the reputation of the Paris sewers changed dramatically during the mid-nineteenth century.

-I-

During the great building projects of Napoleon III's reign, the Parisian sewer system was expanded throughout the city and constructed large enough to allow men to walk upright and boats or carts to travel its length. Prior to the French Revolution in 1789, Paris had sixteen miles of underground sewers; in 1840 the sewage system measured sixty miles, and its length was eighty-nine miles in 1853. In the years that followed, the system expanded to 480 miles of sewer line. A cholera epidemic and progressive attitudes toward sanitation reform accounted for the rapid building of sewers. But the engineers assigned to the construction improved on the old design of the sewers by increasing their size and concentrating on effective methods for cleaning the underground passages.

Parisian fascination with the engineering feat of the new sewers rivaled that of the ancient Roman statesman, Agrippa, who had traveled the course of the Cloaca Maxima. The government, wishing to promote the utility of sewers, encouraged tours of the facility. After the king of Portugal toured the sewer, it became a tourist attraction for the genteel society. Passengers traveled the tunnels in wagons and boats or on the vehicles designed to clean the sewers. The dark underground passages were lit by lamps, and smell was minimal because of the fast-moving water.

J. Bertrand, author of *Eloge historique d'Eugène Belgrand*, found it difficult to describe the one-hour tour of Parisian sewers. "The charm of a boat excursion in well-diluted sewage does not lend itself to being recreated in the tale one can tell of it." An aristocratic woman added, "The presence of lovely women can add charm to the sewer." The allure of visiting a sewer system firsthand has continued into modern times. Today, tourists visiting Paris can still tour the sewers.

Royal Tails

The nineteenth century produced its share of royal stories on the use of the toilet. Some of them are a bit bizarre.

-I-

The royals fall victim to penis envy apparently in the same manner as do their subjects. A courtier told this story about George IV of England (1820–1830), then Prince of Wales. "One of his [the Prince of Wales's] stories was about the huge size of the penis of one of his royal brothers, a fact that he had discovered one night while riding with him in a carriage. His brother had felt the need to relieve himself. When he did so out the carriage window, the water flowed as from a fountain and the driver urged the horses forward to escape what he thought was a rainstorm!"

-I-

Houses not connected to city drains relied on cesspits. All the problems associated with the pits in previous eras continued during the Victorian age. The accumulation of all that waste had surprising results.

Queen Victoria dominated the age that bears her name. Her personal life and morals became a model for Britain's citizens. By all accounts the marriage of Queen Victoria and Prince Albert proved successful. They were married for twenty years and raised nine children. Prince Albert's untimely death in 1861 from typhoid devastated the queen. For forty years following her husband's death, Queen Victoria insisted on wearing black mourning dresses. To honor his memory, Queen Victoria had statues and memorials built throughout the British Empire. She even went so far as to demand that no changes be made to their living quarters at Windsor Castle after his death, effectively freezing time in the year 1861.

Unfortunately, the order that all remain the same at Windsor extended to the cesspits. No cesspits were emptied in the years following Prince Albert's death. They soon began to spill over. The fifty-three overflowing cesspits contributed to disease and the death of servants and inhabitants of the castle.

-I-

Overflowing cesspits were not uncommon. One evening a lord threw a party for his privileged friends. He stood at the door of his manor house greeting his guests as they arrived. Anticipation ran high as everyone was dressed in evening attire for the festivities. While watching the approach of a horse-drawn carriage, the lord cried in horror as the ground seemed to open and swallow the carriage. The vehicle had crossed over a cesspit, creating a sinkhole. Passengers in the carriage died in the cesspit's muck as the lord stood by helplessly.

-I-

The queen is always the last to know. Shortly after her marriage to Prince Albert, Queen Victoria discovered toilet paper. Before the advent of sewers, raw sewage and toilet paper were disposed of in the river. While visiting Cambridge University in 1843, the queen walked along the banks of the river with Dr. Whewell, master of Trinity College. As the story goes, Queen Victoria glanced over the bridge and asked, "What are those pieces of paper floating down the river?" Not wishing to embarrass the queen, Whewell replied, "Those, ma'am, are notices that bathing is forbidden."

-I-

Innovations in the WC encouraged the planners of Queen Victoria's coronation to install a number of toilets in Westminster Abbey for the huge event. Proud of their modern accommodations, the organizers believed they had solved an old problem associated with a large gathering. Then a skeptic raised his voice. Suppose, during the crowning of the queen, the toilets should be flushed simultaneously. The noise would be deafening and embarrassing. With the help of the royal guards and officials, a test using decibel readers was performed to determine if a "royal flush" would disturb the event. No such disaster would happen. The test revealed the sound of flushing toilets would not reach the hall in which Queen Victoria's coronation was to take place.

In 1850 Queen Victoria had a train car built for her private travel. The royal saloon contained a private washroom, but the "convenience" was hidden inside the sofa. Water closets began to appear in Pullman cars in 1874 and finally reached the third-class cars in the 1880s.

-•I•-

While the Industrial Revolution in Britain ushered in new technologies, not all practices connected with industry were a result of a modern invention. Similar to the practices of the Asians and the ancient Romans, the British collected urine for use in industry. During the eighteenth and nineteenth century, urine was used for bleaching wool. The wool industry was a thriving one in Britain and required large quantities of urine. Mill owners would provide local town residents with large barrels to store their urine. The urine was collected weekly, and the residents were paid for the service. Urine used for the processing of wool was referred to as "lant."

Cleanliness, American Style

The American poet Henry Wadsworth Longfellow (1807–1882) was credited with installing the first water closet in America in his home. However, the United States did not formally begin importing water closets from Britain until the 1870s. Finding the British imports expensive, American manufacturers replaced imported toilets by entering the market with their own. The American versions were initially inferior to those of their cousins. Their designs were crude in comparison to the stylish British porcelain and were initially constructed of earthenware. Porcelain bowls sturdy enough to be incorporated into a water closet proved difficult to develop.

Thomas Maddock from New York was the first American to manufacture the porcelain toilet bowl in the United States. In 1873, he overcame the pitfalls of competing with the higher quality British product by improving his design and producing it more cheaply. To sell his porcelain bowl, Maddock carried his fifty-pound demo in a sack over his shoulder door-to-door in New York City. His perseverance paid off and his business succeeded.

-x-

Flipping through newspapers and catalogs of the late nineteenth century, one realizes that consumerism is not new. Each page of a catalog contains several choices and variations of sanitary devices, more than in the present market. Just as modern technology generates new electronic gadgets only slightly improved from last year's models, trade in toilets at the turn of the century offered the consumer new ways to "lay waste." There was the Syphon Jet Bowl, the Washdown Syphon Bowl, the Combo Hopper and Trap, or the Chemical Indoor Closet.

The Chemical Indoor Closet was billed as the alternative to tramping through the mud to reach the privy. It was guaranteed to "protect the morals of your children." A combination of chemicals and water were required to sanitize the facility and prevent odor. The reliance on chemicals meant a nasty burn for anyone falling in.

Joining the burgeoning sanitation market, Bates Company sold the Leighton Patent Enema Chair in 1846. Designed to replace the syringe, the enema chair looked like an ordinary armchair, but was equipped with an apparatus for injections. The patient remained seated and used any quantity of liquid desired.

Celebrating Cesspits

The vast majority of nineteenth-century Americans placed their trust in the outhouse for their bodily needs because water closets proved too expensive for the average person. Houses were built back-to-back in order to share a cesspit placed on the backyard dividing line. Some families built a cesspit under the house to avoid going outside. Unfortunately, problems with rats, odor, and disease often developed from this convenient arrangement. Catharine Beecher (1800–1878), the sister of the famous author Harriet Beecher Stowe, presented her views on privies in a series of homemaking books. As an advocate of female educational reform, Catharine Beecher believed women needed to be educated in home economy. According to Beecher, women had the most influence in the home. As such, a woman needed to have knowledge on home design, landscape, and cooking in order to provide the best environment for her family.

In her book, *A Treatise on Domestic Economy, for the Use of Young Ladies at Home and at School*, Beecher addresses the problems associated with keeping a clean house. She suggests that two privies be built for the purpose. The privies should be attached to the rear wall of the house. Ventilation could be achieved without attracting flies by installing a window that opened when needed and a spring door that insured the privy stayed sealed. Beecher does not mention why two privies are needed other than for family comfort. However, the presence of two privies is consistent with Victorian sensibilities of separating the sexes in all intimate actions.

-I-

The sanitation problems of New York City during the nineteenth century illustrate the challenges faced by city planners contending with the sudden rise of population due to an influx of immigrants. Conditions deteriorated quickly as the city could not keep up with the amount of human waste produced. Managing garbage and sewage challenged city planners as early as the eighteenth century and reached formidable proportions during the nineteenth century. As with other major cities of the period, New York's problems centered on the collection and disposal of human waste.

New York City passed regulations requiring its citizens to dispose of waste in the city-approved manner. Public law stated that privies be lined with brick, stone, or wood, be built a specific distance from the building, and be emptied and disinfected periodically. Lime was to be sprinkled into the pits to assist in breaking down organic materials. According to the regulations, the inspector of lots could enter a home at any time without notice to perform his duties.

By 1871 legislators mandated a sewer system for the city. To pay for the sewer, a tax was levied on landowners. Many landlords refused to connect their buildings to the sewer. The sewer was viewed by the landlords as a luxury their tenants did not deserve. Thus, the sewer system in New York City grew in piecemeal fashion with the rich neighborhoods being connected first.

-·I·-

Privy pits collected the contents of chamber pots for individual tenement houses and private residences. As the city grew, the pits needed to be emptied fairly often. Nightmen, often called "scavengers," emptied the privy pits into a horse-drawn cart. The contents of a full cart were then dumped into the surrounding waters. Obnoxious smells resulted from this dubious method of dealing with waste. In addition, small ships moored in the harbor fell victim to the weight of dung thrown over the wall. Many ships were sunk as a consequence.

The office of superintendent of scavengers was formed to regulate the nightmen and prevent dumping into the rivers. Finding an alternative location for human waste led New York City managers to New Jersey. Waste recovered from the privies of the Big Apple was transported by boat across the river and dumped in New Jersey.

Still, New Jersey could handle only so much of New York's refuse. Another method of disposal had to be found. Turning to the country that gave them the Statue of Liberty, New York investigated a French method of recycling human waste. *Poudrette*, a composting process, turned France's sewer waste into a rich fertilizer, or nightsoil. Animal manure had long been used as fertilizer, but many objected to human dung being used in the same manner. *Poudrette* was voted down by the city council, who could not overcome the aversion to turning human waste into plant feed. Fortunately technology came to the rescue. Early in the new century, the city sewer system had been connected to the majority of buildings. Shortly afterward, progress in the science of waste treatment enabled New York to continue dumping waste into water systems.

-I-

Working in New York City has always been dangerous. In the 1800s watchmen were hired to protect the nightmen as they traveled the city emptying privies. Many of the nightmen had been mugged while working. Technology slowly rescued the nightman from the other horrors of his job. The scavenger no longer needed to be physically in the privy pit to remove the contents. With the advent of a vacuum that sucked up the dung into a tank, the nightman found his job much less repulsive.

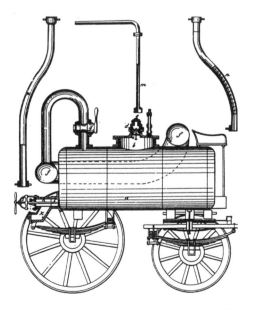

Privy Vacuum patent by F. Daitchy, 1850

—✠—

Excavations of privies are a fascinating way to learn how people lived in the past. Recent excavations of New York City tenement privies reveal that the diet of middle-class New Yorkers in the mid-1800s consisted mainly of stews made with beef, as well as fish, bread, potatoes, and fruit. Popular condiments were English pickles, French olives, and American sauces.

Other secrets divulged by the pit included a tenant who dyed his hair black and another who suffered from a respiratory ailment. Garbage was banned from the privy, but most people ignored the directive and used the pit as a garbage dump. Interestingly, seventeen chamber pots were also found in the pits.

—✠—

The enclosed spaces and flammable contents of the cesspit combined to increase the dangers of explosion. City planners of the growing metropolis were well aware of risk. In 1855, the city of Brooklyn ordered that privies be constructed according to cited building requirements. "No sink, privy or cesspool, shall hereafter be constructed or made within the limits of the lamp and fire districts, unless the same shall be constructed of brick and stone, and be at least ten feet in depth from the surface of the ground when such depth is practicable, under penalty of fifty dollars, to be recovered from the owner and builder of the same, severally and respectively."

—✠—

Before the invention of the earth closet by Reverend Moule, Central Park *was* the "johnny on the job." In the late nineteenth century, to prevent the defilement of Central Park, city administrators ordered over one hundred earth closets.

During the nineteenth century in New York City, the Swamp Angels terrorized the city with their crime sprees. The criminal gang took their name from their modus operandi. They used the sewer system to travel to and escape from the scene.

-I-

A leader in American bathroom sanitary ware, Kohler Company, traces its beginning to the late nineteenth century. John Michael Kohler, an immigrant from Austria, founded the company in 1873. Beginning with an enameled hog scalder–cattle watering trough that could double as a bathtub, Kohler successfully traversed the burgeoning sanitation market. A father of six children (four from his first wife and two from her sister, his second wife), Kohler developed his company into one of the few family-owned businesses still in existence dating from the turn of the century.

-I-

The American spirit is best studied through the settlement of the Old West. The sanitary arrangements in the West testified to the obsession settlers had with making money. Many persons moving west believed they would remain there temporarily, until they got rich. Thus, conveniences such as sewer systems were ignored.

It is difficult to romanticize the Old West in America when you consider the following directions given by a man in Tucson during the 1880s on how to find the governor's house. "You want to find the Governor's? Wa'al, podner, jest keep right down this yere street past the Palace s'loon, till yer gets ter the second manure pile on yer right; then keep to yer left past the post-office, 'n' yer'll see a dead burro in th' middle of th' road, 'n' a mesquite tree 'n yer lef', near a Mexican 'tendajon' (small store), 'n' jes beyond that's the Gov's outfit. Can't miss it . . ." (from *On the Border with Crook*).

-I-

Seattle is proud of its heritage as a timber town in the Old West. The most popular tour in present-day Seattle is the trip through the underground city with an old toilet as the highlight. Initially Seattle was settled on a patch of land below sea level. As the town grew, so did problems with rescuing the dirt streets from floods. Problems of building a city on lowland multiplied. Similar to the monk's latrines built over tidal rivers, the toilets in Seattle were subject to the tides of Puget Sound. High tide caused many of the toilets to explode water three feet into the air.

The attempts made to remedy the problem revealed the stubborn nature of the pioneers. The city constructed paved streets ten feet higher than the previous streets. Shop owners refused to abandon their shops or raise the entrance to street level. Ladders had to be used to cross the street from shop level. Some were thirteen feet high.

High tide in Seattle

⊰ **6** ⊱

The Twentieth Century: Out of the Closet

1920–Present

"Let me say candidly, what comes up must come down. What goes in must come out."

Ching Wah-nan at Hong Kong International Toilet Symposium on Public Toilets, May 1995

Considering the process of natural selection, it was inevitable that the bathroom would gain prominence in the modern household. Unlike the formal dining room, the bathroom tends to be a necessity that cannot easily disappear or be replaced. New house designs reflect the prominence of the bathroom by enlarging it to three times its previous size. Jacuzzis and two-seater bathtubs fit along side the ever-faithful toilet and, in pretentious homes, the bidet. Thus, aside from being a place of leisure, the bathroom makes a statement. Comfort replaces utility, expanse overrides brevity. We've come a long way from the chamber pot placed under a bed.

Although the designs of bathrooms are modern and luxurious, there has been little innovation or upgrading of the toilet bowl. Dull white porcelain of no noticeable design has replaced the artful opulence of Victorian pedestals. Toilets today are designed for water efficiency and ease, not for beauty. It is ironic that the Victorians, with their modesty toward human functions, should have designed ornate and beauteous toilets, while the self-proclaimed, progressive naturalists hide the toilet behind obscure designs. Even outhouses were hidden during the twentieth century. Sunflowers, with their huge stems and wide faces, were planted around the outhouse to prevent knowledge of its presence. Unfortunately, the sunflowers have had the opposite effect. A person looking for relief need only spot the sunflowers to find the outhouse.

The evolution of the modern toilet during the twentieth century presented some creative ideas. After the initial water closet was invented and perfected in Britain, inventors around the world began submitting new patterns, claiming to have made the discovery of the century. A number of these new variations bordered on the bizarre. Although many of the inventions never made it to the production line, the efforts exemplify the creative spirit of modern inventors. Yet, one is left to wonder why the toilet still looks the same after two hundred years. Surely there is room for improvement.

Out in the Outhouse

The outhouse has been around for several hundred years in a variety of forms. Older Americans recall the importance of the outhouse in their lives. They speak fondly of splinters in their butts and smells that could kill a cat. Those were the good old days. The backyard privy still graces the rolling hills of rural America. Whether a hundred yards from the house, built into a corner of the house, or next to the barn or garden, every outhouse has a few basic essentials. A pipe for ventilation rises from the roof, small cutouts—typically a crescent moon and star—let in light and air, and a door usually opens inward to give the person inside control. Beyond the basics, outhouses provided the owner with the chance to expand according to his needs. They were referred to by the number of seats available: "three-holer," for example. Some outhouses resembled the main house. They contained painted wood, etched friezes, and windows with curtains.

There was even a mobile privy. Called a "buck and railer," this primitive privy was built on timbers and consisted of an ordinary board with a railing to prevent falling backwards. The board had no holes. Balance was needed in maneuvering one's butt off the board without slipping into no man's land. Farmers relied on the "buck and railer" for their needs while in the field.

Some outhouses were decorated with framed pictures on the wall, baskets for waste, or a bucket holding corn cobs for wiping. In his book *Backyard Classic: An Adventure in Nostalgia* (or an adventure in nausea), Lambert Florin described outhouses decorated with carpet, wallpaper, mirrors, gold-plated door knobs, and elaborate lids to cover the hole.

However, the majority of outdoor privies offered little enticement for lingering. The air was oppressive, especially in the summer months. During bad weather and nighttime, the chamber pot still served its purpose. Grandma would dump the contents of the previous night in the outhouse every morning. A magnet for flies, most outhouses provided a fly swatter for the occupant, and snakes and rats would occasionally keep the sitter company. To keep the critters to a minimum, a wooden lid covered the sitting hole of the privy.

-I-

Outhouses, called *dunnies*, were a way of life in the outback of Australia. Australians saw the advantage of using the struc-

Servants entrance to privy, William A. Hamill residence,
Georgetown, Colorado (Sandra Wallas/Denver Public Library)

tures for displaying advertisements. After all, the dunnies held
a captive audience, every salesman's dream come true.

Dunnies were built of wood, brick, earth, or grass. Some of
the outhouses provided tin seats rather than wood. The tin
prevented insects and spiders from making themselves a home
as they often did in wood seats. Modesty did not characterize
Australians of the outback. In 1830 a man named William Cox
owned a dunny that provided for nine people at one time.

-•I•-

A man traveling to Norway during the early years of the
twentieth century reported using a precariously located privy.
After breakfasting, he found the outhouse and settled himself
in for a sit. Noticing daylight coming from the hole, he was

astonished to find he was sitting over a cliff two thousand feet above a river.

-x-

During California's gold rush in 1848, cities sprang up overnight throughout the West. A hotel owner in Sierra City, California, serving one of the mining camps appears to have built a privy in a few hours. The two-story outhouse hung dangerously over the Yuba River.

-x-

Outhouses from all over the United States are depicted in Lambert Florin's *Backyard Classic*. He includes a limerick describing the pitfalls of using the "backhouse."

> There was a young man named Hyde
> Who fell into an outhouse and died
>   He had a brother
>   Who fell into another
> And now they're interred side by side.

-x-

Why some outhouses had crescent moons and stars cut into the space above the door is a mystery. The cut-out designs offered relief in the form of ventilation. But some outhouses used pipes rather than cutouts for air. Florin believes that the symbols denoted which sex the privy accommodated. Yet, most backyard privies accommodated both sexes. According to the dictionary of symbols, the moon is associated with the goddess Diana. Diana represented both chastity and birth. When combined with a star, the image denotes "paradise." Possibly the outhouse occupant reached paradise or nirvana as he "gave birth" to his waste.

Country outhouse, Virginia (Julie Horan)

The transition to indoor plumbing occurred slowly in rural America. Eventually the outhouse moved close enough to share a wall with the main house, and then over time it became an "in-house" feature. But before that point, the decorative chamber pot could still be found next to the bed.

A variation on the chamber pot in America was the "slop jar." A bucket with a handle and lid, the slop jar was more attractive than its name would imply. The advantage of the slop jar was the addition of a fitted lid that minimized both odor and the chance of spillage.

Outhouses were found in cities as well as on farms. Nineteenth-century New York City depended on outhouses to accommodate the tenement slums, schools, and prisons. Called "school sinks" because they were designed for "schools" of people, these outhouses were connected to the city sewers and could be flushed with water occasionally. Rarely were they cleaned.

Paying Homage to the Privy

> "The Country Seat," by Christopher Curtis

They were built like emporia in the reign of Victoria
In the castle, manor, or grange,
With their seats made of wood, which have gamely withstood
Pressures greater than mere winds of change

Boys with bats, balls, or oars, Sportsmen sporting twelve-bores
Gaze in rows from the walls on the sitter
Draughts and damp old stone tiles means today's stately piles
Suffer badly from cold that is bitter

But, forget all the strain, pull the gleaming brass chain
(with a porcelain handle, no less)
And, released by a piston, from within a vast cistern
Comes a roar—and you're flushed with success

The Patience of Patents

The original bathroom was located in the bedroom. Water closets commonly connected to sewer lines or cesspits beneath the house, replacing the chamber pot in the bedroom. By the twentieth century, water closets finally graduated to their own room. Barely large enough to hold the toilet, the small bathroom prevented much maneuvering by the user.

Marketing the toilet around the turn of the century required the ability to convince the public of the efficiency of your product without offending "delicate" sensibilities. Advertisements relied on strong, reliable names such as Primo, Renaissance, or Empire. Patents had no need for ambiguity. The patents submitted for toilet designs and other sanitation devices over the past century are curious examples of the inventive mind.

-I-

Improvement in Deodorizing Excrements (1871) Pierre Nicolas Goux of Paris, France, designed a facility that collects dung and converts it to manure within the same receptacle. "I employ a peculiar system of manure-producing closet, cesspool, or receptacle, in which I effect the immediate and complete absorption of the gases and liquids contained in the fecal matters by means of absorbent substances." Pierre Goux creates a haunting image of a "manure-producing" toilet bursting at the seems.

-I-

Urinal Mat (1880) That annoying feature of dripping urine in the men's restroom was solved by Edward J. Mallett Jr.'s invention. "My invention is directed as a means for catching and deodorizing and disinfecting the drippings of urine which, when the ordinary urinal is used, are apt to fall on the floor." Basically, Mallett invented a grate that covered an absorbing substance. Of course, if men could aim there would be no need for the Urinal Mat.

-I-

Privy-Stool (1898) William Bliss of Constantinople, Turkey, found a way to combine the Western preference for sitting and the Eastern manner of squatting to find relief. He describes his invention as "a privy-stool constructed to permit persons to squat thereon, and thus avoid personal contact with the frequently foul and sometimes infected seats, the invention at the same time permitting the stool to be used in the ordinary manner, if desired." Foot pads on either side of the privy seat require a person to be trained in yoga in order to straddle the toilet.

-I-

Flushable Squat Closet (1910)   A popular theme, the combination squat/sit toilets continued to be the subject of patents into the 1930s. Apparently, the American public was not convinced of the merits of squatting in the Eastern way. Mr. Fuller, the inventor of the Flushable Squat Closet, tried to argue the simplicity of his invention. He wrote, "I contemplate the employment of a foot support or supports so located with reference to the feces receiving surface that the body of the user when crouched upon the said support or supports may be properly presented with regard to the said surface."

-I-

Sanitary Closet (1912)   James H. King improved the commode by adding a sliding trap door similar to the flat blades magicians use when cutting their assistants in half. The trap door slides over the hole to prevent odor and disease when the closet is not in use. To protect the rear, King added an air-cushion seat. An occupant would proceed with caution: a drawback of the device was the danger of dangling body parts being severed by the sliding trap door.

-I-

Sanitary Bowl With Hydraulic Drier (1932)   Inventors love to design gadgets that can do several things at once. A favorite device during the early twentieth century was the combination of a toilet and a bidet. What man could resist a gadget with multiple functions, and what woman could resist a two-for-one sale?

Introducing his device with the declaration of "I, Giovanni Battista Menghi, an Italian citizen, do hereby declare the invention, for which I pray that a patent may be granted to me, and the method by which it is to be performed, to be particularly described in and by the following statement." Menghi goes on to describe a toilet bowl with a pipe that swings out pumping water on the user's private parts. After the water, a stream of air dries the area.

-I-

Liquid-Supply Device for Toilet Purposes (1925)   Variations of the genital-cleaning toilet included Charles Dionne's form of liquid toilet paper described as a "means for supplying a cleansing liquid to a wiping or rubbing element for application to the body for cleansing purposes after excretion, and further for supplying water to a wiping element for the purpose of removing the cleansing fluid from the body."

-I-

Hand Bidet (1908)   Other forays into the wet method of cleaning the bottom included the Hand Bidet by F. Weidl. It resembled an accordion designed to splash the private parts by contracting its body.

-I-

Sponge Pad (1930)   The Sponge Pad by Rosco C. Zuckerman was designed to supplement the use of toilet paper. A moist pad that does not disintegrate immediately, it was used to "properly wash and sterilize . . . outside and adjacent parts of the excretory organs of the human body."

-I-

Water Closet (1917)   Luther F. Erwin's *Water Closet* design was one of the most bizarre sanitation devices I ever came across. Indulging in a bit of "circus fantasy," Erwin contrived a way to avoid sitting on the toilet seat. The victim can balance himself on two vertically projecting arms using foot rests, or he can give up and sit forward of the arms.

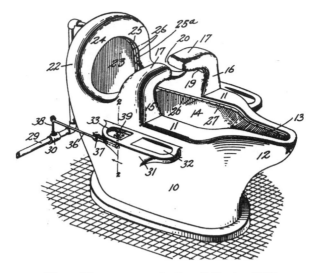

Water Closet patent, Luther F. Erwin, 1917

Method and Means for Operating a Toilet in a Fallout Shelter (1961)   The concerns and paranoias of a society can be evaluated through its toilet designs. Odor, flies, and disease were the concerns at the turn of the century. But by 1961, fears had turned to the effect of the atomic bomb on toilets. Robert O'Brien and Kenneth Milette explored the challenges posed by installing a toilet in a bomb shelter. The goal was to minimize the amount of water consumed. The environment of a shelter required that a toilet operate without electricity, produce little accumulation of sewage, and limit the amount of odor released into the living space. The authors predicted that a supply of fifty to seventy-five gallons of water would last about two weeks in the shelter. The water was used for consumption as well as personal hygiene. O'Brien and Milette's invention used the waste water produced from showers and the collection of urine to flush the shelter's toilet. Fortunately, the urine was also deodorized. The toilet operated by a manual pump that created enough velocity to flush one quart of water, cleaning

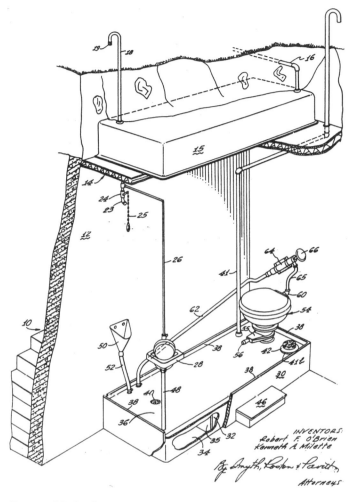

Patent: Method and Means for Operating a Toilet in a Fallout
Shelter, Robert O'Brien and Kenneth Milette, 1961

the bowl. The contents of the toilet bowl were stored in a
closed sewage tank.

The drawing of the fallout shelter confused me. It appeared
that the water source came from a hose perhaps connected to
the house above ground. Also an air vent reached above
ground.

Device for the Collection and Processing of Stool Specimens (1978) Muhammad Javed Aslam of Canada invented a device that purées stools for collection purposes. A disposable contraption is placed in the center of a toilet bowl. At the base of the bowl is a large blade. After the stool is collected and a lid placed on top, the device is placed on a blender base. The tube can be inserted into the bowl's side for extracting the contents without opening the lid. Any kitchen blender can do the same job for less money.

-·I·-

Method for Monitoring the Discharge of Liquids and Apparatus Therefor (1991) The Germans are famous for their high-tech industrial machinery. Toilets are no exception. Wolfgang Lehmann invented a toilet that measured the amount of radiation collected in the toilet bowl after a person has peed. His design was intended for workers in nuclear plants. If the urine discharged is normal, it is flushed. Urine showing dangerous levels of radiation was prevented from leaving the bowl and transferred by an individual to a special holding tank.

-·I·-

Toilet Device With System for Inspecting Health Conditions (Yoshiki Hiruta, 1992) If you ever wake up in the morning and wonder about the content of your urine, this toilet can give you an answer. The Japanese developed a toilet that is equipped with a reservoir to collect and analyze urine. The display panel on the wall records the results.

-·I·-

Public Lavatory System (1981) Struggling to move their toilets into the twentieth century and away from a reliance on the hole in the floor, an inventor from the Soviet Union developed a device that is almost modern. The almost squat toilet enables the user to partially squat over a hole in the floor with the aid of a back support and foot stopper. The device is advertised as perfect for public restrooms because of the hygienic avoidance of body parts touching any surfaces. Of course, the condition of a person's clothes after using this toilet remains to be seen . . . or inhaled.

—I—

Americans are producing patents too. But the American patents are . . . well, see for yourself.

Device for Lowering Toilet Seats (1989)   Sometimes the simplest inventions are the most successful. Timothy C. Probasco connects the flush handle with a device to lower the toilet seat. As the person flushes the toilet, the seat automatically closes. This invention is a must for any woman living with a man.

Toilet Training Device and Method of Use (1992)   Moreno J. White Jr. developed a paper that will not disintegrate immediately in water but will break up once exposed to urine.

Illuminated Commode Training Kit (1990)   James M. Sanders had children in mind when he invented this. Finding the toilet in the dark can be difficult, so Sanders placed footsteps on a mat that glow in the dark, showing the way to the bathroom. A glow-in-the-dark ring surrounds the toilet, marking its presence as well as that of the toilet paper holder. Although meant for children, the device works equally well for people who have had too much alcohol.

Toilet Seat Clock Apparatus (1993)   No American bathroom would be complete without a clock to remind us that time is money. Ron Alsip designed a clock that attaches to the toilet, allowing the user the luxury of looking down to see how late he is for that meeting.

—I—

George Welliever is an example of American ingenuity at its best. George invented a combination toilet and microwave.

Understanding the Technology

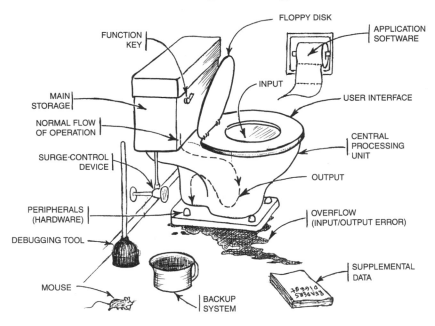

The toilet as a computer

Billed as a water-saving device, the microwaves zap turds into ashes. A similar device, called the Shit Zapper, was proposed by inventor Burt Axelrod.

Green Toilets

The green movement has targeted toilets as an area where improvements would benefit the environment. No longer are toilets manufactured that use three to seven gallons of water per flush. A recent federal policy requires new residential toilets to use only 1.6 gallons. New York City alone could save ninety million gallons of water a day if everyone used the new water-saving toilets. To encourage nonresidential buildings and hotels to make the switch to 1.6-gallon toilets, the New York City government offered $150 for replaced old toilets.

However, there are some problems with the new toilets, namely "skid marks." The low-flush devices do not always provide enough pressure and water to empty the entire bowl. If the bowl does empty, there are often telltale marks left behind. Select-A-Flush has developed a toilet to "dispose" of this troubling problem. The company created a toilet that allows the user to select the amount of water needed to flush the bowl. A pilot project conducted in three Houston apartment complexes found that twenty percent was saved on the water bill with use of the modified toilet.

A cheap method of saving water employs the combination of sink and toilet. Real Goods, an environmental catalog, sells for $35 a sink that replaces the toilet-tank lid and uses clean water from the spigot to flush the contents of the bowl. Thus, water used for washing hands is recycled into the toilet for use in flushing.

A man from Berlin concerned with saving water recycled his bath water by running the water through houseplants to filter harmful bacteria. The water was then used for flushing the toilet. The water ran out of the window through the plants on the side of the building and into a holding tank on the bottom floor.

-◦I◦-

You can recycle just about anything nowadays. Farmers have known for centuries the benefits of composting dung to derive a nitrogen-rich nightsoil called "black gold." Better than chemical fertilizers, nightsoil offers an inexpensive organic fertilizer for crops and overprocessed earth. Prior to 1982, seventy-five percent of excrement was buried in landfills, dumped in the ocean, or burned. Now, more than fifty percent of the offensive matter is recycled. Fear of disease limits the use of the "black gold" to fertilizing crops that feed animals.

But many sewage systems are experimenting with expanding the soil's market.

A contemporary Swedish toilet cuts out the middleman in composting dung. The environmentally correct toilet operates without water or a hookup to the sewer system. Excrement is placed in a tightly sealed packet until bacteria breaks the matter down to its organic contents of nitrogen and phosphorus. It is an improvement on the *Clivus* invented in the 1930s by Swedish art teacher, Rikard Lindstrom. Lindstrom's *Clivus* collected all the waste produced in the house and made nightsoil using the compost process. Only a small amount of water was allowed. Today's version separates urine from the dung for a more efficient system.

Toilet Tidbits

Inventions are only part of the story surrounding the twentieth century's toilets. Anecdotes best describe the effects of modern toilet technology on people's lives. Some are attracted to the electronic devices, and others rely on good old-fashioned waste management.

-I-

Fatherly advice is passed from generation to generation in most families. In the British royal family, Edward VIII, famous for his abdication of the throne, remembered the advice of his father. "Never to refuse an invitation to take the weight off my feet and to seize every opportunity I could to relieve myself."

-I-

The Japanese have produced a majority of today's electronic devices. It is no surprise that they have lent their expertise to the comforts of the bathroom. Similar to the historic Shogun, who didn't lift a finger when using the toilet, today's Japanese can purchase a toilet that takes care of all their needs. On the wall next to the toilet is a control panel. The toilet seat is first warmed for the user at the push of a button. When finished evacuating his bowels, the user pushes another button, causing the toilet bowl to emit a rush of warm water on the buttocks to clean the area. Next, hot air drys the rump. No toilet paper needed.

-I-

Technology is making people lazy. Coffee makers with timers brew the morning coffee before a person wakes up. Remote switches turn on stereos, lights, and televisions as people sit comfortably on the couch. Now the lazy person's technology has been extended to the toilet. A person need not go to the trouble of taking a paper seat cover from the dispenser in a public toilet before entering the stall. Electronic toilet seats will slide out a plastic cover, encasing the toilet seat at the press of a button.

-I-

Writing on his years as an Indian agent for the Canadian government, George H. Gooderham described the process of Native Americans' accepting the sanitation practices of the West. Prior to life on the reservation, Indians had the vast territory at their disposal. The nomadic tribes encountered few problems using streams or shallow pits for human waste. Customs are slow to change. At the turn of the century, the Indians were still using the ground for disposal. Gooderham tells the story of an Indian woman who excused herself from a conversation to wander a number of yards away and squat in a field.

Native Americans were at the bottom of the list for government building contracts. Houses were available for the Indians, but outhouses or toilets were nonexistent. To help the local Native Americans, a carpenter agreed to build several inexpensive, simple outhouses (two-seaters). The Indians came to value the outhouses as if they were gold and guarded them to prevent theft.

-I-

The outhouse received its name from its location in reference to the house. Located at a safe distance from the house to prevent well contamination, the facility could be difficult to find at night. Before the days of electricity, a night in the country was pitch dark. For help in locating the outhouse in a hurry, homeowners connected a long rope or string from the house to the outhouse. To reach the outhouse during the night, one need only to grab on to the string and follow it. An added benefit of the arrangement was the possibility to use the string to hang clothes for drying.

-I-

A microbiologist named Dr. Gerba has made a lifetime study of germs in the bathroom. His expert advice is to choose the first stall in a public bathroom. Gerba surveyed the comings and goings of people and found the majority chose the middle stall for use. Of course after hearing this report, the first stall will no longer be a good choice. Try the last stall.

-I-

Drivers in Bangkok, Thailand, and cabbies in New York City have more in common with each other than just gridlock traffic. The male drivers in both cities depend on portable "urinals" for use while sitting in traffic. The population of Bangkok has exploded in recent years causing horrendous traffic jams. Gas stations in Bangkok sell small red urinals to customers in need of relief during rush hour. In New York City there are no public bathrooms. The city closed all the public restrooms because of their use by drug addicts and the homeless. Cab drivers have to resort to carrying around a glass jar in their cars for "coffee breaks."

⊰ 7 ⊱

Outrageous Hygiene Habits and Customs Around the World

A cultural history of the toilet would be incomplete if it did not include curious habits associated with human waste. Captain John G. Bourke, the American traveler in 1891, collected a striking history of scatological rites. His observations, along with the writings of famous explorers such as Captain Sir Richard Francis Burton and Amerigo Vespucci, shed light on a strange, but fascinating, topic.

Cleanliness means different things to different people. Most agree that some amount of cleanliness is important. But reasons for maintaining a level of sanitation and the form it takes vary according to religion, geographic location, social status, and period of history.

Cultural hygiene habits have long frustrated and mystified the outside observer. Explorers such as Christopher Columbus and Marco Polo recorded in amazement and apprehension the practices of the people they encountered. Ignorance of the logic behind the rituals of foreign cultures led to feelings of prejudice and superiority on the part of the Western colonists in the eighteenth and nineteenth centuries.

Modern technology has done little to inform the masses, although telephones, faxes, and airplanes have made other people more accessible. Discussing toilet facilities with a young Moroccan, the young man stared in disbelief upon learning that Americans do not wash their bodies after *caca*. In the tradition of the French, Moroccans wipe themselves and use the bidet to cleanse after defecating. While traveling in the United States, the young Moroccan tries to time his bowel movements to coincide with his daily shower. We have a lot to learn about each other. The following is a review of past and present customs among many of the world's cultures.

Hindus

Christians have always claimed "cleanliness is next to godliness." A clean body reflects a pure soul. What happens if your religion worships multiple gods? In the case of the Hindus, cleanliness becomes defined by the priestly Brahmins. Brahmins follow the ancient scriptures, the Bhagavad Gita, as a guide to correct living. These holy teachers believe the body is corrupted by the functions that produce disgusting bodily fluids. Thus, constant absolution is necessary to "rectify the rectum." For the common man, absolution comes in the form of thousands of people filling the Ganges River to wash their sins away on a holy day. The Brahmins follow a strict regimen of hygiene according to their spiritual dogma.

How a Brahmin Takes a Dump   The Hindu religion is full of ancient traditions related to the individual's position in society. Brahmins sit at the pinnacle of society, having inherited the responsibility of performing religious rituals. Certain rituals demand daily observance by the Brahmins. Defecating contains several steps according to prescribed ritual. In fact, going to the bathroom probably consumes a large part of the day for the holy man.

First the Brahmin carries his brass pot filled with water to a spot several yards from his domicile. The spot is designated for the purpose at hand. On arrival, he removes his slippers and places them at a distance. After choosing a clean place, he hangs the cord of his wrap over his left ear. Next he covers his head with his loincloth so that he cannot see. The holy man stoops low to the ground. Now comes the hard part. The Brahmin must not:

• Look at the sun, moon, stars, fire, another Brahmin, a temple, or a sacred temple (explaining the loincloth over his face).
• Chew anything in his mouth or hold anything on his head (other than the loin cloth).
• Look at the results of his actions.

In addition, as if there were not enough to remember, the Brahmin must not wear new clothes when voiding. He is forbidden to stoop in a temple, at the edge of a pond, a well, a public thoroughfare, on light-colored soil, a plowed field, or near the sacred banyan tree. Neither standing nor squatting halfway is permitted.

While the Brahmin is dumping he must remain silent. No grunting. When finished, his feet and hands must be washed in the same spot. Not to be forgotten are the private parts. He takes his member in his left hand and pours water over it with his right hand. Next the Brahmin must cleanse his entire body. At the edge of a river or pond, he collects a clod of earth (again there is a long list of types of earth to avoid) using it to clean first his offensive parts, then his hands (left first), then the feet (right first), and the rest of the body. Each area must be cleansed five times with the dirt and then once with water.

The unlucky soul must then wash and rinse his mouth eight times, spitting the water to his left side. Thinking of Vishnu three times, he swallows a little water each time. The number of mouth washes depends on the offense. Urinating requires four washes, defecating eight, eating twelve, and sex sixteen. Any deviance from the above plan will ensure a trip to hell for the holy man.

-·I·-

The Worm Farm   Marco Polo wrote that the Brahmins in India would spread their excrement over a wide area. Believing in the sanctity of life, the Brahmins were afraid their poop would give rise to worms. The worms might die in the desert for lack of food. By spreading the dung in a shear manner, worms were not likely to "grow."

Muslims

Mohammed received divine inspiration when writing the Koran, the holy scripture of Islam. The Koran highlights the dos and don'ts of using the bathroom. Cleanliness was not an issue taken lightly by the Prophet. Followers of Allah were told explicitly what was expected of them.

Muslim Habits   According to author Jcesph Pitton de Tournefort (1656–1708) from his *Voyage au Levant*, Muslim men squat when urinating to avoid contaminating their clothing. After urinating they wipe their penis with one to three portions of stone, clay, or dirt. The ceremonial cleaning must be performed before the Muslim can pray.

-·I·-

Western Reactions   Ironically, early Western travelers to the Middle East remarked with distaste on the clean nature of the Muslims. As reported in *Cleanliness and Godliness*, Henry Blount (1634) derisively described the Turks as believing that "he or she who bathe not twice or thrice a week are held nasty: every time they make water, or other unclean exercise of nature, they wash those parts little regarding who stands by."

-I-

Cruel Joke   Sir Richard Burton also reported seeing Muslim men rub their penises on a stone or with a clod of earth after urinating as part of a cleansing ritual. He said Muslim men carried clods of earth, sand, or stone under their turbans for just such a purpose. A British doctor had his "colonial" fun by rubbing red pepper on a wall frequented by Muslim men. The painful results left the victims running to the same doctor for a cure. In jest, the doctor explained he must cut off the offending part. Having received a good laugh at the expense of his victims, the doctor would wrap the penis in ointment and send the man on his way.

-I-

Cleanliness Is Not Next to the Toilet Seat   Some Muslims believe that it is wrong to sit on a toilet seat that has been occupied by other people. Squat toilets avoid the need to touch a surface. But when a Muslim must use a Western-style toilet, he will stand on the toilet seat and squat to avoid physical contact.

-I-

Left Foot First   Similar to the Hindus, Muslims are required to adhere to certain religious rites concerning cleanliness. Absolution after evacuation of bodily fluids is necessary according to the Koran. The Muslim uses only his left hand to clean himself. Water, or sand if water is unavailable, is required for the cleaning. The bathroom is considered dirty to the Muslim. When entering the room, he must step in with his left foot first. . . . *"And that's what it's all about."*

Questionable Customs

The Role of Urine

'Tis an old receiv'd opinion, That if two doe piss together they shall quarrell.

Ancient civilizations designed various apparatus to dispose of the waste of cities. But the truly ingenious civilizations found ways to recycle the waste. Urine, in particular, had its uses—as bleach, curative agent, or talisman in religious rituals.

The Many Uses of Human Urine

• The Romans bleached their tunics with human urine. As mentioned earlier, Emperor Vespasian placed a tax on the lucrative business of collecting urine.

• Herodotus (d. 423 B.C.), known as the Father of History, believed the urine of a newly "deflowered" girl was ideal for curing medical problems in the eyes. He advocated splashing the eyes with the urine.

• Pliny (d. 79 A.D.), the Latin scholar, found peeing on your feet each morning contributed to one's general good health.

• Residents of medieval Spain used urine to clean their teeth. They believed that urine would not only whiten their teeth but also prevent the loss of teeth.

• Native American Indians, Eskimos, and tribes in Siberia reportedly used urine for tanning and curing animal skins.

• In Ireland, Scotland, and parts of Scandinavia, ammonia derived from stale urine helped in the process of making blue, violet, and bluish red dye from lichens.

• Urine seasoned the food of Indians in Bogotá, Columbia. Replacing salt, human urine was mixed with palm scrapings to create a tasty additive.

• Eskimos' cleaned their hair with urine, while Mexicans found it a cure for dandruff. The Nuer tribe in Ethiopia wash themselves with urine.

• Urine has been known to remove ink stains and make dyes for tattooing when mixed with coal dust.

• Human urine was used to make a fertilizer called *urate* by farmers in Flanders, Switzerland, and France.

• Finally, the urine of eunuchs was thought to make a barren woman fertile. Evidently, the women drank the liquid.

-I-

A secret medicine order of the Zuni Indians, the Nehue Cue, according to Bourke, performed the "urine dance" in which they drank up to two gallons of human urine. The dance was said to teach fortitude to the men as well as cure stomach disorders.

-I-

During the Yin Dynasty in China (1154–1122 B.C.), a concoction of urine was considered an excellent sex drug. Known as "the hunting lion," the aphrodisiac contained the paws of bears simmered over a slow fire and was flavored with the horn of a rhinoceros. The formula was then distilled in human urine.

-I-

In seventeenth-century Europe, many people believed witches who placed spells could be discovered by baking a cake with the urine of the victim. The witch would be compelled to appear to claim the cake. A variation of the superstition has the victim piss into a bottle containing three nails. It was believed that the practice would torment the witch casting the spell by preventing her from urinating. Another way to outwit the witch was to spit into the chamber pot after urinating.

In Germany, as in France, men blamed their impotency on evil spells. The man was advised to form a circle with his finger and thumb and urinate through it. (The circle was viewed as a safe symbol in witch stories; for example, draw a circle at the crossroads for protection.)

-I-

Many superstitions believed urine contained special elements that could protect humans. In eighteenth-century Scotland, the lady of the house would sprinkle her family with urine as everyone awoke on New Year's Day. Midwives sprinkled the birthing-bed with urine before attending to a birth in order to bring good luck.

-I-

Be careful of a Siberian Chukchi offering you a drink. It may be urine from a woman in the house. Custom requires the man to offer his wife for sex to a visitor, who must first drink the woman's urine to prove his worthiness.

-I-

Nomads have been known to slit a vein on their horse to drink the blood when liquids are unavailable. The Qedar tribe in the Middle East and the Apache Indians of North America went a step further. The Qedar killed their camels and drank the urine found in the bladder. The Apaches also drank from the horse's bladder in a dire emergency.

-I-

According to Venerable Bede, the high king of Ireland believed his good fortunes came from having drunk the urine from the privy of a holy cleric.

-I-

Puduer, an ancient Greek writer, believed that women peed according to the tides of the moon.

Finding a Use for Dung

Dung also filled many purposes in history. Human dung was used as a fertilizer in most agricultural societies. However, other less known uses were made of the waste. During the Middle Ages, women used a potion called "love philter-aphilter" to cause an intended victim to fall hopelessly in love. The potion was made of the woman's own excrement. A remedy to the love potion was concocted of human skull, coral, verbing flowers, afterbirth, and urine. As testament to the believed powers of the potion, the use of love philter-aphilter was punishable by death.

-·I·-

Australian aborigines believed witches, *bengals,* used dung to gain power over the person who had defecated the sample. To thwart the *bengals,* it was necessary to bury all dung. The Dyaks of Borneo followed a similar practice. Across the globe in Brazil, members of a tribe living along the Orinoco River were known to carry a hoe with them in order to bury their dung.

-·I·-

During the Christmas holidays, lovers try to catch each other standing under the mistletoe and steal a kiss. In German, the term *mist* means dung. A folktale tells of mistletoe's origin coming from a specific bird that leaves its dung on the end of the tree. From the bird's remains, the mistletoe grows. In many European cultures mistletoe represents fertility, love, and, of course, dung.

-·I·-

The consumption of dung took on ritual tones in the Feast of Fools in medieval Europe, which was thought to have descended from pagan times. Actors, dressed as women or clowns, would eat sausages and blood puddings during high mass at the altar. Linguistic records suggest the sausages replaced human waste as the ritual meal. *Boudin,* the French term for blood sausages, also meant excrement. The Feast of Fools was abolished by King Henry VIII in England, but it did not disappear in France until the French Revolution.

-·I·-

Waste-eating rituals are thought to derive from incidents in history when a group of people were besieged and cut off from food and water. To survive they resorted to eating their own waste. In the Bible, 2 Kings xviii 27, and again in Isaiah xxxvi 12, it is said, "But Rabshakeh said unto them; hath my master sent to the men which sit on the wall, that they may eat their own dung and drink their own piss with you?" The Bible is referring to the Jews who continually fought Roman occupation during biblical years. In several instances, they were forced into an area and isolated for months.

—✠—

I have already mentioned the Roman god of dung, Stercutius, but other cultures represent human excrement as connected to the supernatural as well. The Mexicans have a goddess of dung, Tlacolteotl. She presides over love and carnal pleasures as well as fertility. Ancient Jews identified their god of dung as Beelzeboul, possibly related to the god of ordure, Baal-Peor. The distinction between the two gods is that ordure signifies human dung as opposed to animal dung. Deities dedicated to excrement are common in ancient cultures concerned with agriculture because of the value placed on dung as a fertilizer.

—✠—

Sick people living in the seventeenth century relied on an herb hung in a bag around their neck to ward off illness. Called the "devils dung," the foul-smelling herb was successful in warding off other people from the contagious person.

—✠—

Western tradition has widows in mourning wear black dresses. A similar practice was found in an Australian tribe. Older women in mourning wore human excrement on their heads to show their sorrow.

A Few Thoughts on Sanitation, Hygiene, and Health

Before the advent of modern sewers, toilets, the Clean Water Act, antibiotics, and enlightened physicians, the human race made do with nature's elements to cure the ailments brought on by a deficient sanitation system. Surprisingly, the writers of the past sometimes had great insight into the mysteries of common illness. Nonetheless, few records remain describing treatments for "sickness of the bowel."

-I-

The Friar Is Appalled Friar William de Rubraquis, a Franciscan monk acting as emissary for King Louis IX of France to the Grand Khan of Tartary in 1235 A.D., was not amused by the manners of the Tartarns. "Having list at any time to ease themselves, the filthy louses had not the manners to withdraw themselves further from use than a Beane can be cast. Yea, like vile slovens, they would lay their tails in our presence, while they were yet talking with us."

-I-

Exorcising Fevers Early settlers to America faced illness on a daily basis. Fevers were the most common ailment and required double treatments of bloodletting and purging. Leeches aided in releasing the malady from the sick person's veins. Various herbal concoctions acted like laxatives to purge the other end of the body. Thus, the sick person spent many hours in the privy. Seventeenth-century doctors believed human waste was corrosive to the body of the patient. Therefore, the body had to be rid of the contents of the bowels before medicine could be given.

-I-

Leonardo Speaks The man who invented the helicopter and painted the Mona Lisa wrote widely on various subjects including health and hygiene. His advice: "Be temperate with wine, take a little frequently, but not at other than the proper mealtimes, nor on an empty stomach; neither protract nor delay [the visit to] the privy."

Da Vinci also had suggestions for uses of dung. Dry dung mixed with crushed olives set afire could be used to efficiently asphyxiate an enemy garrison. Dung mixed with ashes aids in the binding of lead with another metal when used as a coating. Leonardo Da Vinci was insatiably curious. Through him we have enjoyed beautiful art and thoughtful philosophy. However, there are some subjects on which ignorance is bliss.

-I-

The Prophet Speaks   "My bowels shall sound like an harp." (Isaiah xvi.ii)

-I-

Ancient Irish Cures   From the land famous for its beer drinks come remedies for curing dysentery and diarrhea. A connection is not made between drinking and diarrhea, but the medicines suggested might lead some to search out a beer instead.

For dysentery Irish physicians believed beer, slowly heated to a boil with a red hot iron rod, eased the bowels. Half of a pint was to be taken in the morning and the other half in the evening. Diarrhea, the "flux of the belly," was cured by boiling sorrel (sour herbs) with red wine and drinking it in small doses.

-I-

Persian Etiquette   According to a British traveler in Persia, the Persians will not pray before a privy or a room that contains a chamber pot.

-I-

Bathing Eastern Siberian Style   Natives in eastern Siberia collect the urine produced by the family in barrels. The urine is used in bathing. The scum that accumulates on the rim of the barrel is used as an insect repellent by smearing it on the body.

-I-

Euro-African Relations   The natives of Guinea were disgusted by the habits of the Dutch when they first encountered them in the sixteenth century. The people of Guinea viewed the flatulence of the Dutch as a direct insult to them.

-I-

Privy Spirits   In the province of Bengal, natives believed that privies were home to the *patni*, evil spirits that took the form of a female with long black hair and club feet. The evil spirits were known to strangle visitors to the privy late at night.

-I-

Chinese Customs   The Chinese living in Indonesia have discontinued many of their old customs in an effort to assimilate to their new home. One of the customs cast off included a superstition surrounding death. The tradition calls for a house not to be cleaned until the body of the dead person is buried. A house may be swept, but the filth must not leave its premises. The belief is that, if filth leaves the house before the body is buried, fortune is thrown out too.

-I-

Jealousy Rears More Than an Ugly Head   There is a psychological theory that, as men cannot become pregnant, they revere their achievement at producing feces. In parts of Europe, toilets are designed with wide bowls that catch the waste as it falls. Before it is flushed away the owner can examine his offering. The anthropologist Margaret Mead considered defecation in the primitive societies she studied to be tied to societal feelings of procreation.

-I-

African Bad Manners   According to Bourke's research in *Scatologic Rites of All Nations*, native Angolans refused to use latrines for their needs. To Angolans, it was considered bad taste to use the same place to defecate. Instead, they would relieve themselves behind a different bush each time, not bothering to cover the dumpings. If a man were to defecate inside the house, he would be ridiculed by his tribe and labeled a *D'Kombe*, leopard.

-*I*-

The Chinese Cane   The Chinese of the past proved themselves to be more refined than most. A traveler described the practice of urinating among the dignified classes: "It is usual for the princes, and even the people, to make water standing. Persons of dignity, as well as the vice-kings, and the principal officers, have gilded canes, a cubit long, which are bored through, and these they use as often as they make water, standing upright all the time; and by this means the tube carries the water to a good distance from them. They are of opinion that all pains in the kidneys ... and even the stone, are caused by making water in a sitting posture; and that the reins cannot free themselves absolutely of these humors but by standing to evacuate; and that thus this posture contributes exceedingly to the preservation of health." The Chinese blamed kidney problems on urinating while squatting.

-*I*-

Picking a Position   The position-assumed while peeing varied among many cultures of the past. The Apache men squatted while urinating and the women stood up.

-*I*-

A Roman Bouquet   Roman women from the upper class drank turpentine so that their urine would smell of roses. No mention was made as to how many women died from the practice.

-*I*-

Manners Among the Walruses   An adventurer in 1565, Dittmar Bleecken, described the table manners of his hosts while exploring Iceland and Greenland. "Neither is it lawful for any one to rise from the table to make water; but for this purpose the daughter of the house, or another maid or woman, attendeth always at the table, watchfull if any one beckon to them; to him that beckoneth shee gives the chamber-pot under the table with her owne hands; the rest in the meanwhile grunt like swine least any noise bee heard. The water being poured out, hee washeth the bason, and offereth his services to him that is willing; and he is accounteth uncivill who abhorreth this fashion."

-I-

The Enema From the Sea   Pliny reported that the ancient Romans liked to use sea water for their enemas.

-I-

Japanese Flower   The Japanese planted a tree called *nanten* close to their privies. Superstition believed that a fictitious animal called *Baku* would approach the privy to eat the red berries produced by the tree. *Baku* was believed to possess the ability to make bad dreams disappear. Thus, those sitting on the pot when *Baku* arrived were ensured dreams that were sweet (even if the air was not).

Modern Japanese children continue to equate the smell of trees with the bathroom. Likewise, bathroom air fresheners disperse the fresh scent of trees.

⋇ **8** ⋇

What's in a Name?

How NOT to Announce Your Intention of Using the Toilet

Euphemisms allow a person to excuse himself politely when in need of urinating or defecating. The Victorians were obsessively modest in avoiding the mention of bodily functions and became professionals in the development of euphemistic phrases. Prior to the prudes of the 1800s, the genteel society relied on common euphemisms such as "going to pick a rose," but Victorian and twentieth-century sayings offer a creativity and humor found in no other period.

⋅—I—⋅

Going to Spend a Penny   During the Great Exhibition at the Crystal Palace, George Jennings provided the use of his improved Water Closet for the public—at a price. The use of the facilities cost each person one penny. Hence, the expression "going to spend a penny" became a popular saying for evacuating one's bowels.

—✠—

The Original John   The origin of "john" as a name for the toilet first appeared in the early twentieth century. However, other forms of the proper name, John, have long been associated with the privy. From approximately 1530 until 1750, "Jake's" referred to the privy. John Harington's book, *The Metamorphosis of Ajax*, probably received its name, Ajax, from a slang of "Jake's place." By the 1770s, Jake had been replaced by "Jack." The diarist, James Woodforde, made the following entry in his journal on January 25, 1779: "Busy this morning cleaning my Jack, and did it completely."

—✠—

King Arthur's Excuse   A saying popular during a period when damsels in distress could be found in castles was, "Within yon tower, there is a flower, that holds my hart." This is a poetic way of excusing oneself to use the garderobe. The term *hart* can be substituted for fart.

—✠—

The British are not the only prudes in the world. A rich tradition of toilet euphemisms can be found in every society.

Chinese

Let Go of Your Hands   During the Ming Dynasty, prisoners were sent to the border regions as punishment. Fearing the prisoners would escape, soldiers tied their hands together. As the "need" arose, prisoners asked for their hands to be freed. The phrase "let go of your hands" became synonymous with stating the need to urinate.

-I-

Yee Ha! For most Americans, the expression "yee ha" brings to mind a cowboy swinging his hat in the air and yelling in delight. To the Chinese the term *yee ha* is used to excuse oneself to use the bathroom.

-I-

Number One . . .   Different cultures can share similar popular expressions. Americans and Chinese both refer to peeing as "doing number one" and defecating as "number two." Referring to bodily functions in terms of numbers was first popularized in the late nineteenth century among English-speaking children. The common use of numbers for expressing the need to use the bathroom should make traveling in China somewhat easier.

Other sayings that will get you to the bathroom while in China include:

"I'm going to a remarkable meeting of philosophers and friends;"

"I'm going to the hall of brotherhood;"

"Heaven grant happiness."

General Toilet Expressions

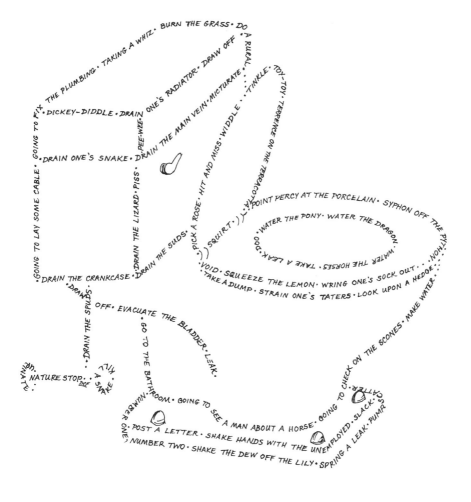

⚜ **9** ⚜

The Roll of Toilet Paper

Before "squeezably soft" Charmin, any small object fell to use for cleaning oneself. Campers can attest to the resourcefulness people develop when in desperate need of a wipe. Leaves are an obvious choice when in the woods. But stones, twigs, and the hand, historically, have all been used in times of need. The evolution of toilet paper developed along some odd avenues.

The Mysteries of Paper

Tracing the evolution of toilet paper, lovingly referred to as TP by many Americans, requires a trip back to ancient history. The Egyptians first discovered that papyrus grass could be made into paper in the year 2500 B.C. In China a court official, Cai Lun, discovered that paper could be made from trees in 105 A.D. The road for paper to enter the West was slow and circuitous. The Arabs gained knowledge of papermaking from the Chinese in 712 A.D. Spain inherited papermaking skills from the Arab occupation in 1150. England finally received the technology in 1590 and America in 1698. After such a long enterprise, it is not surprising that paper was a precious commodity. Had the West not come into contact with the Arabs, we might still be wiping our butts with rocks and leaves.

Although the first reference to toilet paper was in 1718, it probably referred to use in dressing and grooming oneself. No one would dare to think of themselves using such an important article lightly. Paper specifically made for use in the privy first appeared in the United States in 1857 from Joseph C. Gayetty, a manufacturer in New Jersey. Sold as "Gayetty's Medicated Paper—a perfectly pure article for the toilet and for the prevention of piles," Gayetty's product was made of manila hemp paper. Hardly reassuring to think of wiping the sensitive part of your body with the same material as rope.

From the Start

The Economical Choice   Primitive man used the most obvious device to be found when cleaning himself after using the bathroom—his hand. Today many people in parts of the Middle East and Asia continue to wipe with their left hand. Disease is prevented because of the taboo of accepting food offered with the left hand. The people of India take the precaution of sitting on their left hand while eating a meal. If you're not sure which hand to use when tending to the needs of the body, a common rule of thumb can be applied to any situation. The body is divided into spheres. From the belly button and above the right hand is used, for example eating or combing hair. From the belly button and below the left hand is used for cleaning oneself.

A danger in the Muslim culture is the price one pays for stealing. Stealing results in the cutting off of the right hand. An incompetent thief can starve without the use of his right hand. The prophets designed a very effective and far-reaching system of punishment in the "eye for an eye" approach.

-I-

A Desperate Man   The great Roman Empire consisted of more than pyromaniac, violin-playing emperors and all-you-can-eat banquets. Conquering neighboring lands during their many military campaigns, Romans captured men and women from Europe, Asia Minor, and Northern Africa. The hapless souls were brought to Rome and sold into slavery as servants, farm hands, or, in some cases, to provide theater for the citizens of Rome. Slaves trained as gladiators fought each other or wild animals to the death entertaining the Romans. Seneca (d. 65 A.D.), a Roman philosopher, described the desperate suicide of a German slave in an effort to avoid being a candidate for the Colosseum ring. Rather than face death at the hands of lions, bears, or tigers, the slave took the opportunity of committing suicide while using the privy. He crammed the sponge on a stick down his throat, suffocating to death.

-I-

Monkish Ways   Lacking the variety and creativity of the Romans, people during the Middle Ages made due with what was readily available. Wiping, if done at all, was with the assistance of old torn cloth. Archeological excavations have revealed that monks used strips of old clerical robes in the latrine. Considering the immense problem monks faced with constipation and the rough wool of clerical robes, it is a safe bet that they not only had a bald spot on their head but a red, sore, bare spot on their bum. I imagine the prayers of these men often pleaded deliverance from their "pain in the ass."

-I-

The Goose-Feather Problem   The period of the Enlightenment proved that the indulgences of royalty knew no bounds. Believing their butts were more sensitive than those belonging

The goose feather as a wipe

to commoners, they used silk strips or soft goose feathers to wipe themselves. The use of goose feathers presented a slight problem, as they were not stiff enough for wiping. The royals succeeded in solving the dilemma. They left the feathers attached to the goose's neck in order to get leverage.

Victorian Modesty As might be expected, toilet paper posed a dilemma for the prudish Victorians. Toilet paper reminded genteel citizens of the base needs of the body. The invention of the water closet guaranteed privacy. In reaction to the delicacy of the Victorian citizen, toilet paper was sold from behind the drugstore counter to avoid offending sensitive minds. According to Scott Paper Co., women would ask for toilet paper by brand name or lower their eyes and say, "I'll have one of *those*, please."

Double Duty Early toilet paper was packaged in small squares and was not readily available when traveling.

Responding to the market, a device was sold during the eighteenth and nineteenth centuries allowing upper-class women who could afford the tissue to carry it discreetly with them. Madame's Double Utility Fan contained a secret compartment in the handle to hold single sheets of toilet tissue. A precursor to the famous gadgets of James Bond, the Utility Fan enabled the lady to remain cool as she wiped herself.

-I-

A Precious Commodity   The early twentieth century viewed the new and useful privy paper as a commodity in bathroom apparel. As such, toilet paper was guarded diligently. Hotel clerks handed out TP to their patrons in specified amounts. Even today, places in Europe control the distribution of toilet paper. A large woman sits in the restroom selling the tiny, rough sheets to wary tourists.

-I-

Modern Wipes   Rear ends in the pre-TP era were chafed and ink-stained. People must have sent up prayers of thanks when the perforated toilet paper roll came into being. Prior to perforated TP, a knife had to be kept in a cabinet next to the roll of toilet paper to separate the paper for use. By 1880 the British Patent Perforated Paper Company and Philadelphia's Scott Brothers were selling this vital addition to the bathroom. Toilet paper had come a long way from the first leaf.

A Proud Ancestry and Short History

Soft Corn   The United States developed traditions of their own in wiping, namely the corn cob. Not just a joke, country folks were known to use corn cobs much the same way as the Romans used sea sponges. Generally, but not always, the corn cob was soaked to soften it. The cobs were placed in a corner box for convenience. To brighten up the outhouse, corn cobs were dyed different colors much as modern toilet paper is today.

The torture of that icy seat could make a Spartan sob,
For needs must scrape the gooseflesh with a lacerating cob.
Yet the corn cob was not as rough as we would imagine.
Fresh cobs were pliable and often
Came in the different colors.

James Whitcomb Riley

-I-

Toilet Paper American Style   Another form of TP found in America was the Sears and Roebuck catalog. Hanging next to the latrine, the Sears catalog served two purposes—as reading material and as wiping material. So popular was the catalog that when the publishers changed the paper to a glossy type, readers objected, fearing the paper would not absorb as well. As the standard of living increased for Americans in the late nineteenth century, toilet paper replaced the Sears catalog. However, many people in remote rural areas continued to rely on the outhouse and Sears catalog until the mid-1900s.

-I-

Automobile Toilet Paper   The Frantz oil filter offers the consumer a cheap alternative to common automobile oil filters. The Frantz oil filter never has to be changed. Every four thousand miles the owner replaces a roll of toilet paper in the canister and adds a quart of oil.

-I-

Artistic Toilet Paper   Today's toilet paper is soft and comes in a multitude of colors and designs. As a gag gift and novelty item, toilet paper can resemble money, allowing the user to feel rich and frivolous while wiping his butt with government property. Other rolls may provide the user with his horoscope on each sheet. The environmentally conscious wiper can find unbleached recycled TP. In a move that would have shocked Victorians, a company in North Carolina sells a toilet-paper holder that speaks when the paper is rolled. Obnoxiously, it says things such as "yuk-yuk, stinky-stinky, nice one," or sounds alarm bells. What goes around comes around. Early inventors of the toilet worried about silencing the noise to avoid drawing attention to the "necessary" action. Today inventors try their best to announce to the world what is going on.

-I-

Wiping At the Post Office   Writing from London in 1960, Reginald Reynolds suggested that the poor service of the post office in the past could be traced to a lack of toilet paper. "Misplaced" letters had actually been used as toilet paper for the worker needing something with which to wipe in the privy.

-I-

A Survey of Sheets   In 1879, brothers E. Irvin and Clarence R. Scott formed the Scott Paper Company in Philadelphia. In its first years, Scott produced coarse paper goods such as wrapping paper and bags. By the turn of the century, the brothers had expanded into producing and selling toilet tissue. The moral embarrassment caused by toilet paper restricted its being advertised to the public. Instead, Scott sold toilet paper directly to merchants under the private labels of the merchants. In 1902, Scott began advertising their own brand, the Waldorf bathroom tissue. The public had finally come to accept the presence of toilet paper as a part of life.

-I-

As part of a campaign to launch a new brand of toilet paper, the Scott Paper Company conducted a survey among users. Their findings illustrate the manner in which toilet paper defines the person:

• Sixty percent of those surveyed preferred the toilet paper to hang over the top, 29 percent hung their rolls under, and only 7 percent had no preference.

• Two-thirds of people with a master's or doctorate degree reported reading in the bathroom; 56 percent of college graduates and one-half of those with a high school degree admitted to reading on the pot.

• Fifty-four percent of those surveyed reported that they folded the tissue neatly before wiping; 35 percent just wad it in a ball to wipe.

• Twenty-seven percent complained about people who don't replace the roll of toilet paper when it runs out.

• If you are over thirty-five years old you are more likely to tear along the perforated lines. The young continue to live dangerously and tear randomly from the roll (81 percent vs. 52 percent).

• Magazines beat out newspapers for being read more often on the john. Women prefer magazines (56 percent) while men prefer the newspaper (49 percent). Men and women aged eighteen to twenty-four years read the comics (43 percent).

• Only 7 percent admit to stealing toilet tissue from public bathrooms.

The Uses of Toilet Paper

Toilet Paper for the Rich   In Romania the currency has fallen so low that toilet paper was made from five tons of shredded bank notes.

━┰━

Ticker-Tape Parade   Baboons in Africa know how to celebrate. Shortly after Queen Elizabeth II learned of her ascension to the throne while on a safari in Africa, she discovered a tree draped with toilet paper outside her window, as if in celebration. Apparently baboons had stolen the toilet paper from open bathroom windows and had had their fun with it.

-I-

Pakistani Toilet Paper As Muslims, Pakistanis believe in the need to clean oneself with water after defecation. Cleanliness comes in the form of a plastic, brass, or copper vessel with a long curved snout called a *lota*. The most devout have been known to take their *lota* with them when they travel.

-I-

Getting to Know the Ropes Before the invention of paper in China, and in some areas of Africa, ropes were used as a material for wiping after going to the bathroom.

-I-

A Semiprecious Gem The art of making paper traveled to Japan from China in the fourth century but paper was too expensive for use in the bathroom. Two pieces cost as much as a liter of sake. The choice was obvious in terms of enjoyment value. Eventually the price of paper came down enough to encourage people to wipe their butts with it. But it continued for some time to be a luxury of the rich.

-I-

The Russian Way The iron curtain may have come down, but toilet paper is still hard to find in Russia. When available, TP can be bought in stationery stores along with the Mother's Day card you forgot to send. Many Russians prefer to use the daily newspaper. It's cheap and plentiful. They merely cut the newspaper into rectangles and place them in an envelope that hangs on the toilet-room door.

-✕-

The Useful Bidet   An alternative to toilet tissue is the mechanical wonder of the bidet. Many people find the cool, cleansing waters of the bidet more soothing than dry paper. However, American tourists use the bidet for other purposes. It is excellent for cleaning tired feet, hand-washing clothes, or chilling champagne. Bidets also make wonderful plant terrariums.

European bidet (Karn Wong)

⚜ **10** ⚜

The Mysteries of the Orient and Lands Beyond

Although geographically and culturally distant from the West, Asia simultaneously developed strikingly similar methods of handling human waste. Both regions relied on the collection of waste in cesspits or chamber pots. The difference between the two regions lay in the value placed on the dung. Few Western farmers used human waste as fertilizer. Rather, human excrement accumulated in cesspits that were covered over when full, especially in populated areas. In contrast, Asian farmers viewed the dung collected from cities and towns as an important commodity. Perhaps the different views of dung account for the invention of the first modern toilet occurring in the West. A society that abhors natural bodily functions would understandably develop a means to remove the product from sight in a private, efficient manner. Asians are more practical and in tune with nature when it comes to sanitation. They give back to nature its own.

Ancient China

Descriptions of chinese sanitation in the nineteenth century are very similar to Britain and Europe before the advent of water closets and sewers. However, Western travelers relating their experiences in China failed to recognize the similarities and instead reacted with disgust at the less than desirable differences.

Traveling around China during the late nineteenth century, a Westerner described the living conditions as horrendous. "The drainage is naturally bad, and what there is consists of several large canals that penetrate the city from emptying into them; but all these become greatly choked, and, notwithstanding the daily rise and fall of the tide, the city everywhere reeks with odors of the vilest sort, i.e., to Christian noses. The natives do not seem to notice them at all."

As cited in *Scatologic Rites of All Nations*, a writer for the *Chicago News* in 1889 reported that "A traveler who lately returned from Peking (Beijing) asserts that there is plenty to smell in that city, but very little to see. . . . The houses are all very low and mean, the streets are wholly unpaved, and are always very muddy and very dusty, and as there are no sewers or cesspools, the filthiness of the town is indescribable."

In general, humans have been unimaginative and uncreative when it comes to the removal of human waste. Roman scavengers collected the city's waste under the cover of night. The same collection process continued in Europe until the development of the sewage system. Likewise, historic China relied on nightmen to clean city streets and pits of human waste. According to Edward S. Morse in "Latrines of the East":

At Shanghai, as one enters the native town he encounters men bearing uncovered buckets upon the ends of a carrying-stick; these are the removers of night-soil, and they have their regular routes through the city. If one follows these scavengers he sees them going to the banks of a canal near by and emptying the buckets with a splash into a long scow, or other kind of boat, which, after being filled, is towed away to the rice-fields in the country. The stuff is often spilled in the water by careless emptying. The canal has no current, at least not enough to disturb the green ooze and sickly yellow condition of the water, which is thick with foulness; yet beside this boat people are dipping up the water for drinking and culinary purposes.

A typical Chinese household owned variously sized buckets for its sanitary needs. Each room had a small basket that was emptied into a large wood bucket with a close fitting top to diminish odors. Bamboo switches stirred the contents of the bucket. No mention was made of the necessity for stirring waste. Daily cleaning of the buckets consisted of scraping them out with shells and drying them by air.

The "cesspot" of the East was a variation of the cesspit found in the West. Chinese cesspots were large earthen jars sunk in the backyard of a house. A small wooden screen was placed in front of the jar. It was not large enough to avert curious eyes; it merely obstructed the view of the pit somewhat. The fired clay of the "cesspot" prevented the seepage of waste into the surrounding ground and water systems. Ashes were spread over the waste layers to absorb odor.

A Western visitor to Canton in 1835 reported the presence of large stoneware vases sunk into the ground along the road for use by travelers. Modern Chinese continue to use earthen pots in the more remote areas. Even in the major cities, the stand-up urinal in men's restrooms are often ceramic jars.

In China, an oblong urinal is called a piss pot, the same terminology Westerners use for chamber pots. However, the oblong shape enabled the piss pot to double as a pillow. Since pillows were little more than wooden blocks, the decorative piss pots of similar size offered a valuable service in the middle of the night.

-I-

In southern China the women continued to use the bucket in the broom closet. The men were relegated to the public

Chinese pillow/piss pot, eighteenth century (E. S. Morse from
Latrines of the East)

privy down the road. The public latrine consisted of a trench
against the wall with two pieces of four-by-six-inch wood
board. The user balanced on the boards, squatting over the
stream below. Wood partitions divided the building, providing
privacy for the men. All went well as long as the stream moved
through the trenches and out the building.

The Tartars were nomads who roamed the remote regions of
China. According to Captain Bourke, the nomadic nature of the
Tartars can be traced to their sanitary beliefs, which also
accounts for the origin of a Tartan curse. "They hold it not
good to abide long in one place, for they will say when they
will curse any of their children, 'I would thou mightest tarry so
long in one place that thou mightest smell thine own dung as
the Christians do'; and this is the greatest curse they have."

Dung in Deng's Time

Observers wonder what the political consequences will be
with the demise of the Communist patriarch Chairman Deng.
But as China struggles to define itself politically, capitalism is
taking a strong hold of the economy. An example of China's
race to become a major player in the world economy are its
efforts to modernize. Of course, this includes toilets. Squat toilets and trenches are found side by side with Western-style sit

toilets. As with everything China does, the modernization of toilets is taken very seriously.

-I-

The modernization of China is most apparent in the upgrading of toilet facilities. In Beijing, the government held a contest to design public-toilet buildings. They received over 340 entries. A young woman won the contest with a building based on Chinese style, including areas for a newspaper stand, phone, and outdoor chairs. The *Beijing Daily* described the event as "A public toilet revolution." The government plans to build thirty pavilions.

-I-

WalMart has yet to set up shop in China. But the concept of discount prices has become common in certain venues. Using some of the public latrines in China requires a fee, but that includes a small amount of toilet paper. For a special few the charge is half price. Reflecting a cultural practice of respect for the elderly, China excludes persons over the age of seventy as well as the handicapped from paying full price for using the public toilet. Members of the armed forces are also included in this special group, reflecting their value in the eyes of the government.

-I-

Despite the conversion to Western-style toilets, tradition continues to rule the lives of the Chinese. This includes a nonchalance in dealing with the by-products of nature's call to mankind. Environmentalists complain of disposable diapers taking up vital space in landfills and failing to decompose at an acceptable rate. Commendably, a majority of Chinese have never used diapers, disposable or cotton. Instead, a toddler is equipped with pants that have a large hole around the private parts. The mother must be aware of her child's schedule of bowel movements. If lucky she can quickly place the child over a vessel to catch the waste. Otherwise, there is a mess to clean up on the floor.

Japan

The Japanese have a rich history of sanitation. Shoguns and royalty boasted elaborate schemes for handling excrement. As one would expect, privies and sanitation systems in Japan's cities were organized and efficient. Catapulted into modern life almost overnight during the early twentieth century, the Japanese responded to the toilet with ingenuity and pragmatism. In fact, the Japanese, along with the French and Germans, are quickly replacing the British and Americans as innovators in sanitary technology.

-I-

During the first century A.D. the Japanese nobility depended on a box for their waste disposal. Twelve-by-fifteen-by-nine inches and decorated elaborately with gold and silver, the box operated as a portable potty for the rich.

Women had difficulty using the box due to the many layers of clothing they were required to wear. To avoid staining her clothing, a woman would accept the help of a servant in removing the underpants; then a T-shaped wood post placed behind the box would hold up, above her waist, the twelve layers of kimonos that she wore. In this precarious position the woman would relieve herself. Another solution was to anticipate the act some fifteen minutes before in order to undress in time.

-I-

A discussion of the history of Asia's sanitary evolution would not be complete without a visit to Japan at the turn of the century. The Japanese outperformed other Asian communities in the quest for a toilet. Every house in Japan had a privy except the very poor. However, the Japanese did not always possess an indoor privy. In prior years the privy was built some distance from the house, preferably over a river. A name for privy in Japanese is *kaha-ya*, or river house, which aptly describes this ancient means of sanitation.

-I-

Japanese privies consisted of a rectangular opening in the floor with a piece of wood sticking up for the user to grasp. While Westerners back in to the toilet, the Japanese face it. Modern toilets in Japan continue to be approached in this manner. The ancient Japanese privy required a person to squat over the hole. Under the rectangular hole lies a large earthen pot or half of an oil barrel to catch the contents raining down from above. The privies in Japanese homes were located in the front of the house and opened out to the street. People walking past in need of the facilities thought nothing of using a stranger's toilet. Many of the privies found in places of business were works of art. Decorative doors of inlaid wood sprouted elaborate designs. It was difficult to identify them as privies. They looked like cabinets or wardrobes.

-I-

The privies in Japan were emptied regularly by men who *paid* for the right. Waste represented a lot of money as fertilizer. The price was so high that poor people could pay their rent with the proceeds of their waste. Highly productive peasants did not even have to pay rent. They just produced the fertilizer.

-I-

While hunting in the country one day, an eighteenth-century shogun warlord had need of the facilities. He refused to use the privy of the peasants. It was not worthy of a man of his stature. The next best accommodation was the peasant's house. The shogun defecated in the middle of the floor in the peasant's house. Rather than get mad, the peasant felt honored. He proudly showed the specimen to all his neighbors, who displayed just the right amount of jealousy.

—I—

Pride in position was prevalent in Japanese aristocratic society during the nineteenth century. One such aristocrat, a princess, was visiting another family to discuss a future match. She became mortified at the thought of using the toilet. Her honor and position placed her above such common facilities. The sea was a toilet worthy of her position. So she jumped into the sea to pee.

—I—

Ancient Kyoto contained the imperial seat of Japan before the capital was moved to present-day Tokyo. A city rich with history, Kyoto displays its heritage through museum palaces, parks, and shrines. While home to the emperor, Kyoto developed a reputation for refinement and grace. It was the cleanest city in eighteenth-century Japan. Buckets available for the public lined the streets. The emphasis on cleanliness helped beautify the city and develop commerce as the contents of the buckets were then sold for fertilizer.

—I—

The women of Kyoto were known for their graceful, sophisticated ways. One story concerned the way in which they urinated. Writing in 1803, Bakin Takizawa reported, "In Kyoto even ladies in high society urinate standing in public. While walking on the street, I have witnessed a lady urinating while she was standing and facing her back to the bucket [remember the Japanese face the toilet]. She seemed to be unashamed of what she was doing and nobody laughed at her."

The technique of urinating while standing without making a mess is a mystery. Maybe the Kyoto women knew a secret about the female anatomy undiscovered by other cultures. The early twentieth century ushered in an era of modern reform. A city council discussed the need to stop female students from voiding while standing. The days of the talented Kyoto women were numbered.

-ɪ-

In 1603 the capital of Japan was moved from Kyoto to Edo (present-day Tokyo). A seventeenth-century castle in Edo was the site of probably the largest toilet in the country. Built with the expensive Japanese cypress, the toilet was 450 square feet. It consisted of a series of wooden boxes. Each box was three feet in length, 1.6 feet in width and 3.3 feet in height. The shogun would climb into a hole on top of the box and do his duty. The box would then be cleaned by servants in the room.

Before attending to the box, however, the servants were required to pamper the shogun. As the master squatted to answer nature's call, one servant would circulate fresh air by fanning him. Another servant had the dubious honor of wiping him. During the winter the toilet was heated with charcoal. Using the toilet was an event for Japan's shogun. Privacy was only for the less fortunate.

-ɪ-

The transfer of power to Edo marked the beginning of the Tokugawa Shogunate, the ruling family of Japan. During the Tokugawa Period (1600–1867) of Japanese history, the shogun's wife enjoyed a toilet equal in size to her husband's. The difference in architecture was that the wife's toilet emptied into an enormous well. It was predicted that the well could hold ten-thousand years of excrement in its depths.

-I-

Edo had none of the sophistication of Kyoto. The women of Edo could not stand and pee. They squatted. The streets of Edo fared worse. While Kyoto was clean and pristine, Edo was dirty and disgusting. People urinated on the streets as there were no public accommodations. Animals added to the chaos. The stench was unbearable.

Modern Japan

Life in Japan for most people is a combination of modern and traditional values and practices as well as superstitions. Family, friends, and travel consume their free time. The marriage of old ways with new is evident in all aspects of life, particularly the toilet. In Japan, everything has its place. Clutter is not tolerated. This orderly existence includes reserving a special pair of slippers for use in the bathroom. Before entering, the Japanese leave house slippers behind and substitute the "toilet" slippers.

-I-

The Japanese have never lost their sense of humor. Americans have adopted many Japanese cartoons: there are the *Teenage Mutant Ninja Turtles, The Mighty Morphin Power Rangers*, and *Speed Racer*. But it's doubtful the television show *Ugo Ugo Ruga* will catch on with the American public. The show has a character called Dr. Puri Puri, or Dr. Stinky. He is a talking turd with heavy eyebrows that pops up in toilets. Not exactly a marketable doll in the West.

-I-

In Japan most toilets are installed in the part of the house facing north. The north has long been associated with foreboding events. From the north comes the bitter cold that makes the winter unbearable. The sick are always quarantined in a room facing north. It is not surprising that the toilet, home to the more mundane personal habits, should be placed facing north.

If misfortune strikes a home, such as an accidental death or "loss of face," a psychic is called in to find the evil sprites. The psychic may suggest that the toilet be moved to the north of the house if one has not already done so. A misplaced toilet can be a dangerous thing to the Japanese.

-·I·-

In 1985, the Japan Toilet Association was founded to promote and inform the public of the many contributions and advancements of the toilet. International conventions have been held in the past few years gathering the great toilet thinkers of the world. Inventors and producers discuss the latest developments in the sanitary industry. The future of the toilet and possible designs receive special acknowledgment by the participants. Ideas from the practical to the bizarre, such as equipping the backseat of automobiles with toilets, are presented during the conference seminars. (Celebrating should include a fifty-flush salute and a "bowl" game complete with toilet parade down main street.)

In Japan, Toto Company controls the market in toilet products. The most popular toilet sold in Japan is the combination toilet-bidet. The coveted features of high-tech toilets are adjustable washer nozzles and deodorizers. The toilet holds an important place in the life of the Japanese.

Other Asian Countries

India

The second largest populated country in the world (800 million), India has yet to come to terms with the abundance of human waste it produces on a daily basis. Modern cities possess accommodations familiar to any visiting Westerner. However, when visiting city slums or traveling in rural areas, one encounters sanitation practices that have not changed in thousands of years. As the host country of one of the oldest religions in the world, Hinduism, India has developed a highly ritualized and organized manner in dealing with waste when the sewer is not available. Yet, as the population increases it is possible India may be consumed by its own waste.

Premodern times in India offered few choices in personal hygiene. The rich could afford a privy built at a distance from the house. However, the poor (and there were many) had to resort to using a bush or "taking a walk in the garden." The garden has been a logical place to deposit one's waste in many cultures. The gardener was bypassed by depositing fertilizer directly on the plants and flowers. Another favorite area was the barnyard.

The great sage of India strove to combat the divisiveness of the caste system during his lifetime. Gandhi broke with ancient tradition by accepting people from the lowest class, the Untouchables, into his entourage. Despite his good intentions, a taboo existed that proved impossible to break. The cleaning of toilets was, and continues to be, viewed by Indians as a job only fit for an Untouchable. Servants will clean any portion of the house except for the toilet. A separate servant, an untouchable, must be hired specifically to clean the offending device.

Travel throughout Asia and in other parts of the world is generally conducted on railways. Trains provide passengers many conveniences, including the use of a toilet. Unlike the airplane, the train does not have a tank to collect waste from the toilet but empties it onto the train track. This is why it is forbidden to use the toilet while it is stopped at the station. As one would expect, in Asia most of the trains contain squat toilets. The movement of the train may lull you to sleep, but it is not conducive to squatting and urinating. Hand-holds assist in the endeavor, but only an octopus can hold on to the train and his clothes at the same time.

Poor natives depend on the bus system to travel in India. A pit stop is required every so often because of the lack of toilet facilities on-board. Being modest, proper, and religious citizens, the men will pee on one side of the road while the women pee on the other.

Rural life in India is rather primitive by most standards. The toilet facilities are generally the river or, for the lucky few, a bucket placed in an enclosed thatched hut. A can of water sits next to the bucket for rinsing. When using the river, modesty demands that people go early before the sun rises or late after the sun sets.

Southeast Asia

Travelers to Southeast Asia during the nineteenth century saw little in the way of modern sanitation facilities. Many villagers continued to relieve themselves the old-fashioned way by using a running stream as a toilet. If water was not available, then a few minutes behind a bush or a hole in the ground would suffice. The few structures designed for the collection of waste did not take advantage of the lay of the land.

Many people in Malaysia lived in huts built on sticks over the water—a reasonable receptacle for human waste. Unfortunately, water was also a source of food (fish), drinking water, a means of cleaning clothes and cooking utensils, and even a place for the burial of family members. Persons dying from disease are believed to have been inhabited by evil spirits and thus were not cremated before being thrown in the river. A small hut constructed similarly to the living quarters became the "thinking man's room," or the bathroom. The bathroom was built behind the main building and connected by a wood walkway. Everything was placed on sticks perched precariously over the water. Inside, the privy had a wood floor with slats placed at a wide distance across the building. The user would simply do his duty while squatting over an opening between the slats. The privy offered privacy from sight but not from sound. Many Southeast Asians did not even have the luxury of partial privacy. They merely built a wood plank to stand on while dumping into the stream.

-**I**-

Modern Burmese cities offer the same accommodations found in many Southeast Asian cities: squat toilets. Water scooped from a bucket is used for flushing the contents down the drain. Because the floor becomes slippery, coconut-leaf stems are used as brooms to brush the water away. However, traveling to small villages, a person will often find only an outhouse, if lucky. The outhouses are located a distance from the village. Some locations don't use a housing structure and only provide an open pit. When using the pit or squatting behind a bush, be sure to cough to let passersby know you are there, thus avoiding intrusions.

Malaysian outhouse (E. S. Morse from *Latrines of the East*)

-I-

The wild jungles of Borneo are famous for their orangutans and headhunters. Primitive tribes living in the forest are surprisingly polite, the shrunken heads hanging from their huts notwithstanding. Living communally in primitive longhouses, tribal women use the kitchen corner for answering the call of nature. The men of the tribe find a bush away from the settlement to attend their needs. While a person is squatting, it is considered ill-mannered to stare or watch the act in progress. Miss Manners would be proud.

Tribes live in communal longhouses—long one-room huts built off the ground on stilts. The more modern longhouse depends on drains in the rear of the house that empty into a cesspit below. Garbage and human waste find their way down the drain. A log on the top of the latrine acts as a toilet, allowing the user the benefit of meditation while eliminating. Balancing his bum on the log, the user has a bird's-eye view of the pigs waiting below for new additions to the cesspit.

A few longhouses have an outhouse located a distance from the house. During the monsoon season, a trip to the toilet can be a death defying act, as flash floods are common.

A few years ago UNICEF gave one of the gateway towns in the rain forest in Borneo one-thousand public toilets in an effort to improve sanitation in the area. The experiment failed. The inhabitants used the privies for a few months, but when no one cleaned them, the natives went back to the old-fashioned way of "going to the bathroom," using the river.

-I-

Romantics have been known to climb to the roof of a building and lay back, watching the stars. In Pakistan many families live on the roof during the summer. The roofs are complete with toilets tucked away behind a brick enclosure. Similar to the garderobes of the medieval castles, the Pakistanian toilet empties the waste down a covered path to the sewer at ground level, an improvement over the moat.

When the British departed present-day Pakistan, they left the "seat of government" behind in the form of a stool with a bucket underneath. The adapted close-stool can still be found in out-of-the-way resorts and homes. Toilet paper is almost impossible to find, but you can be assured of a clean bucket. A man called a sweeper has the job of emptying the bucket after each use.

Finding Relief in Other Regions of the World

Africa

Watch where you step when traveling in Madagascar. Journeying into the island towns presents problems to the weary tourist. It is not unusual to discover both men and women urinating in the streets.

The city of Zanzibar, as described by the famous nineteenth-century British traveler, Richard Burton, conjures thoughts of a pig sty rather than romantic images of Arabian nights. The beaches were littered with feces since people used the sand as a litter box. The water became a cesspool for the entire city. Burton even reported seeing corpses floating in the water. He did not recommend walking the streets of the great city. Garbage and animal and human waste spotted the ground like mini obstacles.

In South Africa many blacks continue to live in townships that lack indoor plumbing, but they usually have a community outhouse. For blacks living in rural South Africa, accommodations are worse. Their houses are built from mud bricks and the floors are made of a mix of cow dung and dirt. When the need arises to relieve themselves, they will walk a distance from the house to squat in the open.

Russia

Communism did little to advance toilet facilities for rural Russians. In remote areas, toilets are still nonexistent. A lucky few have a hole in the floor near an outside wall. In earlier times, villages had a common hole located near a wall by the river. The contents that did not make it into the hole were swept into the river. Winter in Russia is brutal for everyone. But it is particularly hard on the people of Siberia. Russians living in this harsh climate empty their frozen cesspits with pickaxes. One advantage to living in the bitter cold climate: cesspits don't smell during the long winter.

Residents of Russian cities have apartments that divide the bathroom into two compartments. The toilet is in one room and the bathtub and sink in another. Western Europeans have a similar accommodation. The philosophy behind this practice focuses on the belief that using the toilet is considered unclean and requires a room of its own.

South America

The Wai Wai, an indigenous tribe in South American's Guiana region, combine the modern world with their traditional customs. Between their huts and the fields they cultivate is a region known as "no man's land." Villagers are forbidden from defecating in the fields and living area. But within no man's land, privies are provided for sanitation. Despite the presence of the privy, many of the villagers prefer to walk into the rain forest to "do their business" or to squat balancing on a log over open ground.

⚛ **11** ⚛

Private Privy:
The Government and the Toilet

Long before President Lyndon Johnson held meetings with Robert Kennedy while sitting on the john, the toilet played a leading role in governing our nation. Preceding chapters have described the manner in which communities have handled human waste. But how has the United States Federal Government tackled the problems associated with defecation in times of war, peace, and exploration? History tells the story of a move away from using human waste in defense, as in the castle moats of the Middle Ages, to viewing waste as a major enemy to national defense.

The War Between the North and the South

The role human waste played in the Civil War highlights the different resources available to each side. Both sides were concerned with disease among their troops. The filth and disease caused by quartering thousands of men in confined areas competed in horror with the battles of war. A soldier from

Confederate North Carolina wrote, "these Big Battles is not as Bad as the fever . . ."

The North formed the U.S. Sanitary Commission in April 1861 in an effort to prevent disease among its troops. Hygiene habits became a priority. The commission, inspired by the British Sanitary Commission of the Crimean War, adopted the proposed sanitary regulations of leading health experts such as Dr. John Ordronaux in *Hints on Health for the Use of Volunteers* (1861).

According to Dr. John Ordronaux's report, privies, urinals, and cesspools located in the barracks must be continually disinfected unless built over a stream of water. Chloride, sulphate of lime, sulphate of iron, and sulfuric and muriatic acid gas were poured into the pits daily. By preventing the hatching of flies, disease could be eliminated. During the night, urinal cans were fixed near the door of each dormitory to discourage men from urinating on the ground rather than walking to the nearest privy.

Camping in the field presented more of a challenge to hygiene. As a start, privies were set up at the extreme end of the encampment, away from the tents. A reflection of the prevalent Victorian attitude concerning modesty, it was suggested that pits be surrounded by bushes. Each day, the earth collected from digging the pit was used to cover the accumulated waste. Fearful of the sick, a separate pit was dug for "the dejections [excretions] coming from hospital patients."

Other than a surprise attack from the enemy, diarrhea was the greatest threat to a camped army. As a prevention, Dr. Ordronaux advised that soldiers wear flannel next to the skin regardless of weather or time of day. By covering the "abdomen and loins with an apron or bandage," Ordronaux believed the soldier could avoid catching diarrhea and dysentery. Coupled with that unsound advice, Ordronaux recommended that a person with diarrhea abstain from drinking liquids.

The South treated human waste as a valuable asset. A shortage of the chemicals used in making gunpowder threatened to sideline the Rebel cause. By using the nitrate gathered from distilled human urine, the South was able to continue manufacturing gunpowder. Women contributed to the war effort by emptying their chamber pots into collection wagons.

World War I

Military handbooks written in 1915 continued the theme of managing and maintaining sanitation among the ranks. The facilities remained the same although improvements provided more comfort. Depending on the location and length of stay, the field camp had several choices for the disposal of human waste. "Official army business" could be dealt with through burning, burial, or transport by streams.

An unpopular method for the disposal of excrement was the burning of collected waste in a stove built for that purpose. Besides the obvious drawback of odor, burning the waste was

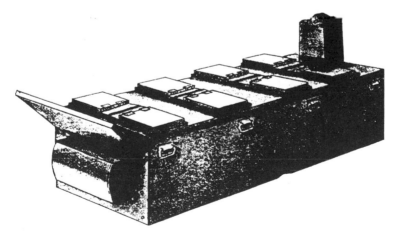

Excrement incinerator (P. M. Ashburn from *The Elements of Military Hygiene*, 1915)

expensive and unmanageable. The equipment proved to be costly and difficult to move. An ideal situation for dung disposal was to make camp by a stream. Water closets could be connected to the stream by a wooden trough, or seats could be built directly over the water. No matter the choice, using the stream for controlling human waste provided a clean and sanitary environment for the soldiers.

-ɪ-

An unlikely method of dumping waste was to carry it away from the camp in pails or carts. Rarely used, collecting and carting away excrement involved a good deal of manpower and time. The majority of military camps relied on pits to meet the sanitary needs of soldiers. Pits were convenient, cheap, and relatively clean if managed properly.

Arriving at the campsite first, the sanitary corps selected a location for the pit on the opposite end from the kitchen. In a temporary camp, the soldiers built a shallow "straddle" pit. For longer stays, the pits were more complex. The straddle pit was only a foot wide, allowing men to straddle the pit when defecating. To accommodate a large garrison, several ditches would be placed side by side. Squatting on the side of a straddle pit was not permitted as the men were known to make a mess dragging their poop into the camp.

Field camps built to last longer than a few days provided wood seats over deep pits for sanitary needs. Eight to ten feet deep and three feet wide, several of the ditches lined the designated area. The number of pits provided had to handle 5 to 8 percent of the company at one time. Different types of seats were used over the deep pits. A common seating arrangement included the use of the Harvard Box, named after its inventor, John Harvard. Consisting of three holes, the Harvard Box resembled the inside of an outhouse but could be carried to the next camp. The no-frills privy seat was a pole fitted between crossed sticks suspended over the pit.

World War I soldier using pit toilet

As the pits filled with excrement to within a foot of the top of the ditch, the pit was covered with dirt and the space marked before a new pit was dug. Burning the contents in the pit afforded another method of disposal. Containing the fire in the pit required the constant attention of an orderly but did not depend on the use of a contraption. The soldier in charge would throw straw and crude petroleum into the pit to start the fire. A stick was used to stir the feces into the fire. Horse manure was an alternative fire-starter. A horse produces eight pounds of manure a day, which can be used to burn the waste of four men. The trick in burning the dung was to get the fire white hot.

Upkeep of military privies was difficult. Soldiers had to be reminded not to throw trash into the latrines, urinals, and privies. The debris, usually clothing and food, clogged the workings of the devices. To prevent obstructions, especially in latrines hooked up to a sewer system, a wire netting or a large, wooden screw was placed at the opening to act as a strainer.

Housing for army latrines depended on the location.

Permanent camps furnished small shelters for water closets and urinals. Shelters were built from galvanized iron, wood, or brick. Many latrines lacked roofs, providing fresh air and a view of the stars.

Discouraging the presence of flies in the latrine proved a challenge. The preferred method appears to have been spreading crude oil on the top of the pit. As backsplashes could be painful to the bum, preventing the oil from splashing was accomplished by a thin strip of tin suspended at an angle to deflect the urine or feces as it fell. Another measure taken against flies included flooding pits with water and lime to kill the larvae. Every two days a pit had to be burned with crude oil.

Latrine pits were used for defecating and urinals for urinating. The urine pit was full of stones covered with grass and straw. The urinal emptied into the a pit either directly or by a slide. Jutting out of the pit were funnels, waist-high for convenience. The soldier peed into the funnel, which was absorbed by the pit. Every so often the funnel needed to be poked with a blade of grass to keep it unclogged and free of flies.

An army of volunteers required special attention and training in hygiene in the 1900s. Prior to battle, men were told to "go to the bathroom." Being hit in the stomach with a cannonball was worse if excrement was present.

It was important for the soldiers to only use toilet paper and not newspaper or other popular wiping materials. Only toilet paper could prevent the anus from being "scratched or irritated." Soldiers were expected to examine their poop closely after relieving themselves. It was considered a good habit for detecting health problems.

Camps and barracks provided soldiers with converted urinals to prevent the men from peeing in any convenient spot. Pails were hung on the wall near the door of the building or barrels placed outside the entry. Men of every culture and period in history have been known to pee wherever they fancy.

The Front Line

World War I changed the rule book of sanitation methods in warfare. Millions of men lived for months at a time in trenches dug into the ground. Living like rats meant living with human waste. Troops on the march or in temporary camps relied on shallow trenches for the disposal of excrement. The use of trenches within trenches seemed a good idea at the beginning of the war. A small trench off the front line trench was designated for receiving the troop's daily givings. The folly of the setup became apparent with the first rainstorm.

Buckets quickly replaced the sanitation trenches. Empty kerosene cans were converted into chamber pots. With one thousand men creating 600 pounds of poop and 300 gallons of pee a day, attending to hygiene needs consumed much of the soldiers' time. An assembly line of men carted full buckets to the back of the line as they were relieved by troops carrying fresh buckets to the front. The waste was buried or burned. A warning was given *not* to dump it in the nearest shell crater. Although convenient, dumping in a shell crater could contaminate the surrounding environment and lead to disease.

As the war dragged on, comfort in any quarter was welcomed. The relief buckets were converted into a close-stool by placing them in boxes with a lid. A soldier could sit and enjoy a little rest between battles while attending to nature's call. Some of the boxes even divided the contents—urine to one side, feces to the other. By separating the contents, a bucket could accommodate more men before being sent to the rear. The buckets were stored in bombproof shelters off the main trench, preventing a potentially *huge* mess. Some trenches included a urine pit comprised of stones and covered with straw. Men could pee directly into these pits, as the "water" would be absorbed quickly.

Soldiers who wandered outside the trenches into the small French villages nearby reported finding antiquated sanitation.

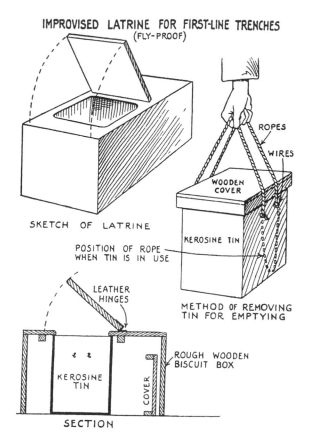

IMPROVISED LATRINE FOR FIRST-LINE TRENCHES
(FLY-PROOF)

ROPES

WIRES

WOODEN
COVER

SKETCH OF LATRINE

KEROSINE TIN

POSITION OF ROPE
WHEN TIN IS IN USE

METHOD OF REMOVING
TIN FOR EMPTYING

LEATHER
HINGES

ROUGH WOODEN
BISCUIT BOX

KEROSINE
TIN

COVER

SECTION

Biscuit bucket converted into a latrine (Joseph H. Ford
from *The Elements of Field Hygiene and Sanitation*)

Water closets were rare. Instead, households used a funnel-shaped opening in the floor that emptied into a pit. Remarkably primitive, the pit was not reinforced with cement, contained no seats, no trap door, and no arrangement for flushing with water.

World War II

Propaganda was transformed into an art form during World War II. Every aspect of daily life in the United States was open to slogans concerning the war effort. Posters, films, radio, and newspapers reported on the evil of our enemies: the Jerrys and the Japs. "Victory Gardens" and "Rosie the Riveter" reminded citizens of their duty to the country. The toilet contributed psychologically to the war effort. In Europe, a chamber pot was manufactured with a picture of Hitler on the inside with the words "Oh, what do I see!" It was a reminder to "score one against him."

Science contributed to the public image of the enemy with a biological theory on the German anatomy. Dr. Edgar Berillon, writing in the *Bulletin of the Society of Medicine of Paris*, described his theory regarding the "excessive defecation" and "body odor" suffered by Germans. He believed the identified bodily functions to be racial defects that could lead Germans to commit unnatural crimes. Dr. Berillon claimed to be able to detect a German spy using urinalysis because, he claimed, Germans possessed 5 percent more non-uric acid than other peoples. In addition, an autopsy of a German revealed the supposed cause of excessive defecation. The Jerry's large intestine was *nine* feet longer than normal. (The doctor added that sometimes the French were known to have this disorder as well.)

During the war in the Pacific, the Japanese captured several American servicemen in the Philippines. The POWs were transported to Japan on "hellships." The ships lived up to their name. On one ship, two thousand American prisoners were packed into a cargo hold. With barely enough room to stand, the men shared food buckets and latrine buckets in an assembly-line style. In the dark, cramped quarters, few could decipher which bucket was food and which was excrement. Within only three days, over 300 POWs died.

The Modern Military

The United States possesses the best-equipped and most well trained military in the world. At a moment's notice, the armed forces can respond to trouble anywhere on the globe with confidence in their ability to meet the challenge. Although the airplanes cost millions of dollars and the bombs are smart, dealing with human waste while in the field has changed little since the turn of the century.

The Army

In 1862 during the Civil War, General McClellan's army was unable to capture Richmond because over 100,000 of his men had diarrhea. Diarrhea claimed the lives of 70,000 soldiers during the course of the war between the states. The majority of victims fighting in the Spanish-American War did not die from bullets but from diarrhea. World War II continued the trend with the fighting force known as Merrill's Marauders in Burma being disbanded because of the toll taken by diarrhea. One platoon had the disease so badly they cut the bottoms out of their pants to expedite baring their bottoms.

All these incidents were cited in a course, prepared in the 1980s and introduced by the United States Army Infantry School, entitled Introduction to Field Sanitation. Sanitation continues to be of paramount priority in modern warfare. The enemy (or enema) is dysentery. The source is flies. The cause is improper disposal of human waste.

Surprisingly, the devices used for disposal of waste while in the field are not modern technological gadgets but the same devices described earlier in the military handbook of 1915. They include straddle trench latrines, deep-pit latrines, burnout latrines, mound latrines, bored-hole latrines, pail latrines, and urine soakage pits. In the absence of plumbing, soldiers in the field rely on burying their dung in the ground. However, today's soldier has a better concept of the consequences of poor hygiene. Close attention to excrement disposal through use of disinfection or burial has prevented the dangerous spread of disease among the ground troops.

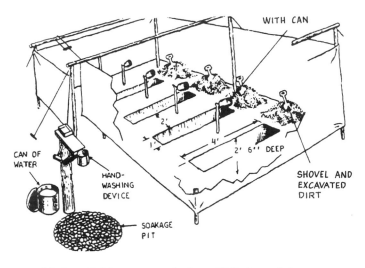

WITH CAN

CAN OF
WATER

HAND-
WASHING
DEVICE

SOAKAGE
PIT

SHOVEL AND
EXCAVATED
DIRT

2' 6'' DEEP

Modern field accommodations (U.S. Army Infantry
School, 1984)

P.C. doesn't only stand for politically correct. It is also the name of a toilet invented by Richard Diaz. The Personal Commode is designed for the soldier in the field. The P.C., a plastic-lined, ten-inch cardboard device, collapses to fit in a backpack. Biodegradable, the liner can be disposed of safely. The P.C. comes in beige or military camouflage.

The Navy

The navy has an easier time dealing with human waste. The sea provides an ideal dumping ground. While in port, waste tanks hold the sewage. Leaving sewage in foreign ports of call is bad diplomacy. Once in the open water, the tanks are emptied of their contents.

On navy ships, keeping the sewage system unobstructed is the job of a group of enlisted men. Not exactly the elite SEALS commandos, the Shit Patrol, as they are affectionately called, perform an important job in the maintenance of the ship. The Shit Patrol walk the stalls of the latrines, unstopping clogged "shitters." A little-known fact about toilets onboard navy ships is that water used to flush the bowls is salt water from the sea.

The presence of women in a historically male bastion created confusion in the armed forces on how to accommodate the opposite sex. Separate living quarters and bathrooms had to be built. Early provisions for women reflected the traditional views held by men. In the 1970s, the navy aircraft carrier, USS *John F. Kennedy*, painted the women's latrine pink while the men's was grey.

The Air Force

Despite the fact that a majority of flight missions last only a few hours, the air force also has problems providing sanitary technology for its pilots. Planes and helicopters possess a urinal tube for the pilot. The device is hidden beneath the pilot's seat. Using it requires a series of awkward movements. Reaching under his seat the pilot retrieves the urinal tubing. While maintaining control of his aircraft, the pilot unzips and removes his member from the full-body flight suit. Then he attaches the device to the protruding member. The pilot is out of luck if he needs to defecate. The arrangement is inadequate for female pilots needing to defecate or urinate in-flight.

The awkward nature of urinating on flight missions results in most pilots waiting until they reach the ground to relieve themselves. If they cannot wait for their destination, some pilots will make a "pit stop" on the way. While flying cross-country, members of the famed Thunderbirds have been known to leave their flight sequence and land at the nearest airfield to use the bathroom.

Tragically, the difficulty of using the urinal tubing while in-flight has led to death for a few pilots. The cockpit cabin of a fighter plane is very cramped. Maneuvering the seat and one's clothing requires the pilot to unbuckle his seat belt. A former F-16 pilot described an incident in which a pilot's seat belt had latched on to the side stick control while he undressed, causing the plane to swerve out of control and crash, killing the pilot. In another incident, a pilot inadvertently ejected himself from the airplane while attempting to pee into the urinal.

Space

In 1961 Alan Shepard made the first successful American manned space flight. Although not the first man in space, the historic event marked the beginning for the United States of manned exploration of the last frontier. Modern man may reach beyond his world to space, but nature will not let him forget his humble beginnings.

Preparing for his legendary flight into space, Shepard practiced daily for months, getting into his pressurized space suit and climbing into the cramped quarters of the capsule. On the day of liftoff, the presence of clouds and some minor mechanical problems left Shepard in the spacecraft for four hours on the launch pad. Suddenly a major problem arose; Shepard had to pee. He couldn't delay the launch and go to the bathroom. It had taken a couple of hours to stuff himself into the bulky suit and attach all the monitors to various parts of his body. Besides, the whole world was watching.

In his book, *The Right Stuff*, Tom Wolfe faithfully records the solution to the crisis. Shepard relayed his situation to ground control. The obvious solution was to allow Shepard to urinate into his suit. The problem, however, was that the monitors attached to Shepard's body were programmed to detect a rise in body temperature and, in response, release freon to cool the body. Urinating would raise the temperature within the pressurized suit. How the suit would respond upon entry into space was not certain. Not wanting to scrub the mission because the astronaut had to use the bathroom, mission control told Alan Shepard to go ahead and pee in his suit.

The relief Shepard felt in his bladder was quickly replaced with concern about the functioning of the pressurized suit. As expected, freon was released, but luckily, it dissipated as the temperature returned to normal. Cradled in the capsule with his feet in the air, the urine slowly crept toward Shepard's head. As the urine made its way up his body, Shepard wondered if it would cause an electrical short in the many wires of his suit. Finally the urine settled in an area on his back. Catastrophe was adverted.

The Apollo missions of the 1960s and 1970s tried to avoid dealing with solid human waste by limiting the fiber in the astronaut's diet. When those moments could no longer be postponed, the Apollo astronauts used a "fecal containment bag" attached to their bodies with adhesive under the pressurized suit. Urinating was much easier. The astronaut placed a condomlike piece on his member that was hooked up to a tube leading to a bottle. The device is similar to that used on military airplanes.

The disposal of solid waste in space proved to be another challenge. In an atmosphere of zero gravity, the poop that collected in the fecal bags often escaped while the astronauts were trying to dispose of it. A few astronauts resorted to beating the flying poop into the bags with a spatula. Once caught, the feces had to be neutralized by mixing it with a disinfectant. The astronaut would knead the contents of the bag like dough till all was mixed.

Flatulence became a larger problem in space than on earth. The genius of the Apollo spacecraft's design allowed it to produce water as a by-product of its fuel cells. The astronauts drank the carbonated water, which was similar to club soda. Another by-product of the design was flatulence. Instead of using thruster rockets to power the spaceship back to earth,

Buzz Aldrin allegedly suggested they use their own jet propulsion.

The Skylab missions presented little improvement in toilet facilities. A marginally successful waste-management system was developed that dumped the collected waste into a tank. A modified toilet with a handhold and a seatbelt was mounted to the wall. Rather than using water to flush waste, engineers designed a system that relied on air. Dubbed the "super john," the toilet worked in zero gravity by first applying a puff of air to the astronaut's bottom to aid in loosening and separating the poop. Then a second blast of air was used to flush the waste into the toilet. This extraordinary toilet also separated the solid waste from the liquid. A drawback to Skylab's system was that disposing of the waste required two astronauts to operate the air-locked tank. One manipulated the device while the other read a nine-step checklist.

The collection and disposal of waste caused many problems for astronauts. Solid matter was stored in waste tanks until it could be properly disposed of on earth, taking up valuable space and weight. Dung was not thrown out into space for fear that it would collide with the ship. Urine was ejected outside the spaceship. However, that disposal method caused more trouble than expected. In space, urine vaporizes into a cloud of fog that follows the spacecraft. To avoid blocking the astronauts' view, the release of the urine was timed to occur after photographs of earth and space had been taken.

The era of the space shuttle marked the final ascent of the toilet in space. But it came with a hefty price. Originally the waste-management system was designed to accommodate a seven-person crew for seven days. Modifications were made as the missions extended to fourteen days. Finally it became apparent that a new system would need to be installed if the shuttle were to continue its groundbreaking explorations in space. NASA contracted for a sewage system from Hamilton Standard. The new toilet made its debut onboard the space shuttle *Endeavor* and was an immediate success with the astronauts. Controversy arose when the price of the new commode was revealed. NASA had paid a whopping 23.4 million for the system.

In defense of the toilet before the congressional subcommittee on space, a spokesperson from NASA described the functions of a waste-collection system on a spacecraft. More than a toilet, the system acts in place of a sewage-treatment plant "operating in a weightless environment, contained in a space one-half the size of a telephone booth, and using the power equivalent of four 100-watt light bulbs." Without the presence of gravity, a continuous stream of air is needed to detach the waste from the body, flush it, and prevent it from floating. In the small confines of the shuttle, with seven people relieving themselves for two weeks or more, the management of waste becomes an important health concern.

The $23.4 million toilet addressed the problems of existing waste systems. It improved the method of storage and compacting of waste while accommodating women in a more comfortable manner than before. In addition, the new toilet can monitor the crew's urine for experimental purposes.

⊰ **12** ⊱

The Last Movement:
Toilet Stories

Here is the scoop.

Fear

Sitting on the toilet with pants around the ankles places an individual in one of the most vulnerable positions known to mankind. Forget about someone mistakenly walking through the door, and instead beware of what may chance into the pot beneath you. Anything can wander into your toilet through the sewer lines. There have been reports of snakes, rats, and squirrels surprising the person lifting the lid.

One such surprise awaited a woman from Georgia. While she was relieving herself, a squirrel appeared and attacked her rear. A tourist traveling in Cambodia found himself in a similar predicament. While using an outhouse located over a cesspit, he was bit on the butt by a pig having some fun in the pit below.

Unfortunately, the FBI does not keep records of violence occurring on the toilet, so it is impossible to assess how many people have been attacked in a like manner.

Wealth

A man from Texas received an unexpected windfall. Over a period of days, seventy-five pens found their way to his toilet through the sewer. The mystery remains unsolved. No one at the sewer company can explain how the pens got there. Attesting to their durability, the ballpoint pens still write.

-I-

In the quest for a money-making venture, an entrepreneur devised an aid for the lazy dog-owner. Doggy toilets will make walking the dog obsolete. The invention adapts the average toilet to accommodate a dog. Stairs are placed beside the toilet, allowing the dog access to a plank placed over the bowl. The dog merely squats while balancing himself on the plank.

Black Comedy

Be careful where you place your baby. A woman traveling by train in Zimbabwe took her baby with her to the toilet. Unexplainably, the baby slipped through the toilet hole onto the train tracks below. Two newscasters were suspended for laughing uncontrollably while reporting the story. Not a word was said about the fate of the infant.

-I-

The explosive nature of flatulence is known to many children who have experimented by placing a flame close to a fart, producing a small explosion. The same event occurs in sewer systems caused by the buildup of methane gas. Recently in Ohio, theory became fact. Backed-up sewer lines released powerful methane gas during high winds, resulting in a toilet's catching fire.

Evidently, exploding toilets are not that rare. A family in the Netherlands had their toilet explode three times in about two years. The Dutch man complained, "It destroyed the whole pot." The cause of the explosion was thought to be pollution in the ground.

-I-

"The Collapse of Toilets in Glasgow" was reported in the December 1993 *Scottish Medical Journal*. Three people were injured when the toilets they sat on collapsed unexpectedly. Rushed to the hospital, the victims received medical attention for the wounds received on their butts.

-I-

The public works division is responsible for keeping the city's infrastructure running smoothly. What happens when a small detail is overlooked during the construction of the division's own new building? In Yakima, Washington, the toilets explode. It seems the small detail that was forgotten was to hook up the building's sewer lines to the main sewer that transports the waste to the treatment plant. As a consequence, the sewer lines filled until there was no room left. Then all the toilets exploded.

Crime

There is not much a person can do to defend himself or herself when sitting on the commode with pants around the ankles or dress above the waist. Thieves have capitalized on this fact, snatching purses hung on the door while women use the facilities. To combat the rising incidence of purse snatching in New Jersey public bathrooms, the police had all the hooks removed from the stall doors. The crooks knew a good thing when they saw it, so they merely took it upon themselves to replace the hooks and continued to steal.

-I-

In Vermont, a landlord got a big surprise when he saw the recently burned remains of the house he had rented to a family for twelve years. Health officials wearing protective clothing went in to investigate and found approximately eighty five-gallon buckets of human feces in the home. Apparently the tenants had refused to arrange for plumbing repairs, relying instead on the old-fashioned way of waste disposal.

-I-

While David Dinkins served as mayor of New York City, there was a rash of thefts in government buildings. The objects of desire: toilet-bowl handles. Over 109 toilet-bowl handles were stolen—including one down the hall from the mayor. No one knows why.

-I-

Mismanagement may not be a crime, but the results can be. The performance of Washington, D.C.'s municipal government hit an all-time low in May 1995. The city's supplier of bathroom products stopped future deliveries, citing Washington's failure to pay its past bills. Thousands of city workers were forced to use bathrooms with *no* toilet paper.

Special Effects

Recycling at its best has reached New Delhi, India. Collecting waste from over forty public toilets, and processing it to provide energy, resulted in the creation of lighting for streets and gas for cooking. Given the 800 million people in India, energy could be produced for the entire country, maybe even the continent.

-I-

Mary Roach interviewed Dr. Chuck Gerba, a microbiologist who specializes in studying toilets, and wrote an informative article titled "What I Learned From Dr. Clean . . . " What I learned from the article was enough to make the hygiene-paranoid Howard Hughes commit suicide.

Dr. Gerba, in his considerable investigation, discovered a phenomenon in the toilet he calls "aerosolization." Every time a toilet is flushed, a cloud of invisible spray is released into the air containing all the bacteria found in a toilet bowl. That spray settles on every surface within six feet, including your toothbrush and hand soap. Closing the toilet lid does not solve the problem because the aerosol lasts for two to four hours and will be released into the face of the next person who opens the lid.

How did Dr. Gerba verify the existence of the toilet spray? He used a procedure called the "commode-a-graph." After spraying blue dye into the toilet bowl, Dr. Gerba placed a piece of paper over the bowl and flushed. The paper collected blue dots, indicating the presence of thousands of gross bacteria.

What can a person do in the presence of germs everywhere? Not much, evidently. True cleaning is not accomplished unless the cleaner is a disinfectant. Gerba, knowing more about toilet hygiene than the common man, cleans his toilet with a fireball. A small amount of alcohol around the rim of the toilet bowl, the touch of fire, and whoosh, a bacterialess toilet. Not a procedure recommended for the average household.

Commercialism

Everyone has enjoyed the ramblings of drunks, crushes of schoolgirls, and prose of budding poets on the walls of public bathrooms. The popularity of bathroom humor is attested to by Madison Avenue's recent line of television commercials featuring references to bodily functions. Now print advertisers want to cash in on the trend by placing advertisements on bathroom walls. Bad idea. Nothing the advertising community comes up with can be as entertaining as the amateur bathroom-wall writer. Of course, we may be going back in time to the days when a *Sears Catalog* hung on a nail next to the commode. Today's advertisements may meet the same fate as the *Catalog*, that is, be used to wipe the butt and flushed down the drain.

-I-

The history of graffiti dates back to the first appearance of walls. Cave walls bear the evidence of the art, or graffiti (depending on the audience), of premodern man. Ancient city walls, such as in Rome and Pompeii, were often covered in political and social graffiti. However, bathroom graffiti has a shorter history. Until the sixteenth century, few buildings had a specific room designated for the toilet. The exception was the Roman public latrine. Graffiti artists were known to leave drawings or phrases on the latrine walls in Rome. However, public officials thwarted their efforts by painting murals of gods on the walls. No one dared deface a god.

In his book, *Graffiti: Two Thousand Years of Wall Writing*, Robert Reisner managed to collect a small sample of graffiti written as long ago as the seventeenth century on old privy walls. The following serve as an example revealing that we are not so different from our ancestors.

England

Privies are now Receptacles of Wit,
And every Fool that hither comes to sh-t,
Affects to write what other Fools have writ.

Privy at Epsom-Wells

Your are eas'd in your Body, and pleas'd in your Mind,
That you leave both a Turd and some Verses behind;
But to me, which is worse, I can't tell, on my Word,
The reading your Verses, or smelling your Turd.

Privy at the Nags' Head Tavern in Bradmere

Central Europe

Upon this board I have put my sheetings,
Whoever sits here has my greetings.

Men's room, Prussia, 1910

Pompeii, 79 A.D.

Apollinaris, doctor to the Emperor Titus, had a crap
here.

Public lavatory

New York City

Here is the place we all must come
To do the work that must be done
Do it quick and do it neat
But please don't do it on the seat.

Anarchists please learn to flush.

Toilet in a university

Japan

"Let's Refrain From Urinating in Public!"

Sign at 1964 Olympics

Twisted Psychology

In Freudian psychology, many personality traits are linked to oral/anal experiences in childhood. The latest twist on Freud's theories explains anxiety levels in sporting games.

According to Itzibi K'Aibozh, a student of psychology and mother of three from the small town of Czipmnink in the former U.S.S.R., young children who throw objects from their cribs are displaying anxiety at fecal loss. She believed that rigid toilet training in the West increases this anxiety in children. They experience horror and stress at the realization that their poop, flushed down the toilet, is gone forever.

Itzibi K'Aibozh offers an alternative to high-pressure toilet training based on the practices of peasants in her home region. Parents allow their children to evacuate their bowels on *durghts* (dirt-floor nurseries). The "natural setting" of the dirt floors is supposed to lessen the child's anxiety about poop disappearing down a drain. Instead, the child watches as his offerings are cleared away by a caretaker.

Sports anxiety enters the picture in the giant leap of supposition that the rigid toilet training of the West will also affect the anxiety level of an athlete. K'Aibozh believes that one-on-one games, such as tennis, remind the athlete of the anxiety experienced during toilet training because they watch the ball disappear. However, in group games such as rugby, athletes do not control the ball individually. Team athletes experience less anxiety because they can comfortably share the ball. K'Aibozh identifies the ball with feces. According to K'Aibozh, athletes involved in team sports experienced little anxiety when they were being toilet trained, whereas individual-sport athletes probably had traumatic experiences with traditional Western toilet-training techniques.

War Between the Sexes

In the March 1993 edition of *Working Woman* magazine, Gail Collins made an astonishingly insightful observation—women may be climbing the corporate ladder, but true success is measured in access to the bathroom.

An obvious example of the theory is found in the halls of the nation's capitol. For years female senators have had to use the same bathrooms as tourists, located yards from the senate chamber. A bathroom for female senators was only finally installed in 1992.

In the Connecticut State capitol, representatives resorted to fighting, not over the budget, but over the bathroom. Women's bathrooms are few and far between, so some of the women commandeered the men's room as unisex. The men retaliated with larger signs of MEN ONLY on the door. At the time the U.S. Congress passed an important bill on equal access, and a new sign appeared on the bathroom door in the Connecticut State capitol: MEN AND HANDICAPPED WOMEN ONLY.

Romance

The lure of romance, travel, and the toilet combine in the excitement of becoming a member of the "mile-high club." The famous club includes members who have had sex in airplane lavatories. Excitement from committing a deviant act is the only reason for choosing an airplane bathroom for having sex. The tiny and usually disgustingly dirty lavatory hardly makes for a romantic rendezvous. A Houston pilot offers a better way to join the club. He takes couples flying in a small jet for romantic encounters.

Down-Home Privy Tales and Others

A farmer complained that his field workers were spending too much time in the outhouse. No amount of scolding deterred the men from their private time of "relaxation." For close to an hour, the men would read the paper while taking care of the morning's biological needs. To deter the lengthy reading session, the farmer used a saw to roughen the edges of the wooden seat in the outhouse.

-ɪ-

An old country woman was heard to remark, "When I was a young girl, I could piss through a wedding ring; now I couldn't hit the side of a cow."

-ɪ-

Sitting in a bar, a man and a woman became competitive after a few rounds of drinks. The woman bragged that she could piss farther than a man. "That's impossible! I'll bet you five dollars I can outpiss you," replied the man. "Okay," answered the woman. "But I want one concession." "Sure," agreed the man knowing he could beat her in any situation. "What is it?" "No hands," the woman answered.

-ɪ-

A Presbyterian newsletter shared the following story with its members. Curious about the toilet facilities at a camp, a woman wrote the camp director and, in her modesty, used the initials B.C. for bathroom commode.

The camp director replied:

Dear Madam,

I am pleased to inform you that a B.C. is located nine miles north of the campground and is capable of seating 250 people at one time. I admit it is quite a distance away if you are in the habit of going regularly, but no doubt you will be pleased to know that a great number of people take their lunches along and make a day of it. The last time my wife and I went was six months ago, and it was so crowded we had to stand up the whole time we were there.

The camp director had interpreted B.C. to mean Baptist Church.

George H. Gooderham, a retired Indian agent on the Blackfoot Reservation, tells the story of how an old man died while sitting in the privy. "An oil company had drilled a deep dry hole in a farmyard and was about to fill it in when the farmer came along and said, 'Don't do that; I'll move the privy over the hole.' The old man settled himself over the freshly dug privy pit for his morning deliberation. When he did not reappear, his family became worried. On checking, they found the old man dead. What caused his death? The grandson remembered the old man saying that when he used the toilet he always held his breath until he heard the plunk!" The hole drilled by the oil company was so deep, the old man had died waiting for the "plunk."

-I-

A gas station attendant must have been bored and looking for some fun when he rigged up a speaker in the privy seat of the women's restroom. As a woman sat down, a voice would announce, "Say, lady, will you move over to the other hole? I'm trying to get this job of painting done down here."

-I-

Lord Byron (1788–1824) was best known for his romantic poetry. But, at Long's Hotel on Bond Street in London he was notorious for his bad hygiene habits. While staying at the hotel, Byron refused to leave the building to use the privy located in the backyard. The night was cold and rainy. So instead, the famed poet defecated in the hall outside his door. Byron was kicked out of the hotel when his action was discovered.

-·I·-

Strolling the streets off the medieval Grand-Place in Brussels, a person will find a small fountain of a boy peeing. The famous Mannekin Pis has a long history in Belgium. Commissioned in 1619, tales surrounding its origin have become legendary. Many variations have been circulated. One story has it that during the battle of Ransbeke, the son of Duke Gottfried of Lorraine was put in his cradle up in a tree to give the soldier courage in fighting. During the battle, the little boy showed his courage by climbing down the tree and urinating. The most popular rendition recalls a little boy pee-

The Mannekin Pis (Julie Horan)

The Jeannekin Pis (Julie Horan)

ing to put out a fire (or bomb fuse) in the town hall to save the city.

Although the tales surrounding the Mannekin Pis are fiction, the statue itself has managed to escape some harrowing experiences. Brussels was bombarded during a war in 1695, but the "little boy" survived, only to be kidnapped by the British in 1745. The statue was returned and stolen again by the French two years later. Louis XV was so upset his countrymen had kidnapped the Mannekin Pis that he arrested the guilty persons, returned the statue, and sent the statue a costume of gold brocade. On holidays or to mark the visit of a dignitary, the Mannekin Pis is clothed in a costume. The City Museum on the Grand-Place holds the 345 costumes and medals worn in the past by the Mannekin Pis. They include an Elvis Las Vegas outfit, several military uniforms, and national costumes from all over the world.

Reflecting the woman's movement of the twentieth century, a Jeannekin Pis fountain was installed in a side street off the square. The Jeannekin Pis models a young girl squatting as she urinates. Unlike the "little boy," the "little girl" does not get dressed up on special occasions.

-I-

According to the Chinese writer, Jiaming Zhu, Joseph Stalin's son proved to be an embarrassment to his father. He died an ignominious death in a German concentration camp during World War II. The young Stalin was held in the concentration camp with a group of British officers. The officers complained to their captor of the barbaric mess created by Stalin when he used the toilet. Young Stalin, furious at the complaint, demanded his honor be restored via punishment of the Brits. The Germans ignored both him and the British officers. Arguments over the condition of toilets were of no concern to the Third Reich. Believing he had been utterly humiliated, Stalin threw himself on the electric wire surrounding the concentration camp and died from electrocution. Zhu commented that the young man died, not in the name of honor, but in the name of shit.

-I-

The explorer Richard Burton led an amazing life. As an adventurer in the nineteenth century, he discovered civilizations unknown to the West. Burton displayed great respect and curiosity for cultures different from his own, and he learned the language and cultural characteristics of his hosts quickly.

Curious to see the holy city of Islam, Burton joined a group of Muslims on their way to Mecca. Only Muslims were allowed to enter the holy city. In order to see the it, Burton donned a disguise of a Muslim. His dark hair and dark eyes enabled him to appear as a Middle Easterner when dressed in traditional robes. His Arabic was flawless and his gestures accurate.

Legend says he made one small mistake during his pilgrimage to Mecca—that could have cost him his life. While relieving himself on the journey, he stood while urinating. Muslims squat to pee. At that moment Burton's companion knew he was not Muslim. Burton killed his companion in order to keep his identity a secret and save his own life. He became the first Westerner to see Mecca.

⊷I⊷

Desert nomads of the Middle East had long depended on nature to assist them in daily living. To maintain sanitation while making camp, the nomads relied on beetles to carry away shit. Similar to the practice in Egypt, the beetles would find the dung, roll it in the sand, and bury it. Western aid organizations hoped to improve the life of the nomads by building them toilets. Unfortunately, modern ways are not always better than traditional methods. The shiny new toilets brought more than sanitation with them. The jewel of the West created a haven for flies, causing the spread of disease.

⊷I⊷

A guide from Elvis's home, Graceland, described the circumstances surrounding the great singer's death. Apparently, Elvis valued the solitude of his private rooms on the second floor. He would often tell his friends not to bother him while he was in that part of the mansion. On one such occasion, his associates began to worry when the "King" did not appear after a few hours. Finally, someone was sent to Elvis's private quarters to enquire. He was found lying on the plush, red carpet next to the toilet, dead. Reportedly, Elvis had suffered a heart attack while relieving himself. Interestingly, many of Elvis's loyal fans wanted to know, "Was he reading the Bible?"

THE TOILET DICTIONARY

Since ancient times, devices for waste disposal have enjoyed more names than shapes. The modern reader can easily be confused by the terms used to describe toilets and their enclosing structures. This short glossary attempts to explain many of these terms encountered in the story of the toilet.

bidet An Old French word that meant a small horse and, as a verb, to trot. By the eighteenth century, bidet was being used to describe a bathroom appliance. Designed to be used to clean the genitals and anal area, especially after defecating, the bidet has gained a tainted reputation.

bucket A bucket or pail worked as well as a chamber pot when in a pinch. In the past, metal buckets were given to prisoners for double duty as a food pail and urinal.

cesspit Also commonly referred to as a cesspool, the cesspit held sewage and garbage. Consisting of a simple hole in the ground, the cesspit was sometimes lined with brick or wood. A lined pit delayed the inevitable seepage of the waste into the surrounding ground. The depth of the pit depended on the wealth of the individual and the type of ground. Shallow pits required constant emptying, which was expensive, while a deep pit needed only be emptied once a year. Other names for the cesspit were privy pit and compost hole.

chaise percée The French term *chaise percée* described a chamber chair or close-stool.

chamber chair Improving on the close-stool, furniture designers in the eighteenth century took an ordinary, stylish chair and converted it into a device that held a chamber pot. The hinged seat opened to an oval seat underneath, usually padded, and in the enclosed depths of the chair lay the chamber pot. The chamber chair was also called the commode chair, necessary chair, or close-stool chair.

chamber pot In one form or another, the chamber pot has existed since man's first permanent dwelling. It is a pot into which one urinates and/or defecates. Rich and poor alike relied on the chamber pot to avoid leaving the comforts of the house to find relief. A simple clay pot or a painted piece of art rich in design with leaves, flowers, or animal motifs, the chamber pot was often found in the bedroom or dining room.

The chamber pot has been known by many names over the centuries. A few of them include: alter, article, bedglass, bishop, cesspot, chamber of commerce, chamber utensil, chamber vessel, chantie, chamber, crock, daisy, gash-bucket, gauge, gut-bucket, guzunder, it, jemima, jerker, jeroboam, jerry, jewel case, jock-um-gage, jordan, jug, lagging-gage, lookingglass, master-can, member mug, mingo, night-glass, potty, pot, pot de chambre, privy pot, and white owl.

cistern   A cistern is a reservoir that primarily holds rainwater. However, some writers use the term interchangeably with cesspit.

close-stool   Dating from the Renaissance, the close-stool resembled a box with a lid opening to reveal a circular seat. Under the seat, a chamber pot was housed to collect the waste. The close-stool proved an improvement on the naked chamber pot. The user could sit for hours on the box rather than squat over a pot. Some close-stools could be locked, preventing anyone but the owner from using it. Unfortunately, not much could be done to cover up the smell. The close-stool eventually developed into the chamber chair, which had arms on the sides. The close stool was also referred to as a commode, night stool, or chamber box. During the nineteenth century, inventive minds designed a close-stool with bed steps.

commode   Another French term, *commode* means convenient—an appropriate word given its function. But *commode* also refers to a chest of drawers dating from the late seventeenth century. A confusing title, the commode might function as a storage chest for household items or, in the eighteenth and nineteenth centuries, to hide a chamber pot. In addition, the commode is a close-stool to some people and a modern toilet to people in the southern United States. Other names for the commode include: stool, Sir Harry, night-stool, pot cupboard, night commode, bedroom commode, or night table.

dung   Dung is another word for excrement, usually referring to the waste of animals, but also to that of humans.

earth closet   The earth closet, invented by Reverend Moule of Britain, operated much like the water closet but with earth. Pulling on the lever released dirt, ash, or sand to cover the poop. Also called the dry privy or the Nettie, the earth closet was popular in areas with a limited water supply.

garderobe   The garderobe was a medieval latrine built into the wall of a castle, usually overhanging a moat or river. It offered a cold reception and no privacy. A garderobe was also known as the privy, necessary, or priest hole.

gongfermor   The gongfermor worked during the night emptying cesspits around medieval cities. Later centuries referred to these sanitation workers as nightsoil men or nightmen. In Japan, the wagon used by the nightmen was called the "honeywagon."

lavatory Although Americans refer to the bathroom as the lavatory, it is actually a free-standing sink for washing one's hands.

latrine Latrine is another modern term for toilet, but it is also used to refer to a primitive toilet such as a wood seat placed over a cesspit.

nightstand The term nightstand was coined in the eighteenth century by Thomas Chippendale for the popular pot cupboard.

night table Another derivative of the close-stool, the night table resembled the commode in many aspects. Popular during the eighteenth century, the night table appeared to be a simple cupboard above a set of drawers. However, while the top of the table held a washbasin and pitcher, the drawers held chamber pot that could be removed and used.

outhouse A popular term in America for the privy, outhouse describes a building over a cesspit usually detached from the main house and used for relieving oneself. Several of the names found under privy also apply to the outhouse.

piss Piss means to urinate or pee.

pissoir A French-term for public latrine, built in the late eighteenth to mid nineteenth century. Popular in European countries.

pot cupboard Specifically designed to hide chamber pots, pot cupboards were popular during the eighteenth century.

privy Privy describes a location or a facility for defecating. When referring to answering nature's call, it can refer to a corner in a medieval castle, a pit in the ground with or without shelter, or a room in a house. The most common definition of a privy is that of a structure built over a cesspit containing one or more seats used for defecating. Multiple seats, poor ventilation, and lots of foul aroma characterize the privy. Some privies eventually moved indoors, being connected to the back of the house. Ironically, the British also use the term for a government office: the privy council.

Nicknames for the privy have become synonymous with those we hear today for the toilet. They include: Ajax, back, backhouse, backy, bank, bog, boggard, bog house, bog shop, cacatorium, can, chapel, chapel of ease, chic sale, closet, coffeehouse, coffee shop, comfort room, comfort station, commode, commons, compost hole, cottage, counting house, crap house, crapper, crappery, crapping casa, crapping case, crapping castle, crapping ken, cropping ken, dilberry creek, donagher, donegan, donigan, draught, dunagan, dunnaken, dunnakin, dunny, dunnyken, forakers, forica, garden house, gong, gong house, honey house, hoosegow, hopper, house of easement, house of office, ivy-covered cottage, Jacques', jake, jake house, john, KYBO, leak house, library, little house, necessary house, necessary vault, office, outdoor plumbing, parliament, petty house, place, place of convenience, place of easement, place of resort, place where you cough, potty, pot, prep chapel, private office, quaker's burying ground, rear, relief station, river house

(Japanese),   Scotch ordinary, shack, shit hole, shit house, shitter, shouse, siege, siege hole, seige house, Sir Harry, smokehouse, Spice Island, stool of ease, temple, Temple of Convenience, The Path, uncle, unflushable, Vandyke, wardrobe.

toilet   Americans use the modern toilet everyday (unless living in the mountains of West Virginia). The majority of modern toilets work with a siphon. The bent pipe, or siphon, allows water to rise up through means of air pressure and over the edge to a lower level. The open end of the siphon is below the water line. Pressing down the lever of the toilet starts the water flowing through the siphon pipe. As the water rounds the bend, the rest of the water in the tank follows due to the air pressure. The motion of water moving through the siphon is stopped when water falls below the level of the open, wide part of the siphon. At the same time the ball falls with the water level, opening the valve to allow water back into the tank. The valve is closed when the ball floats to the top. Simple physics, really.

urinal   Urinals are appliances specifically designed to handle urine. In Europe, urinals are located in enclosures scattered throughout the streets. Dating back to the turn of the century, urinals accommodate only men who can duck into the concrete structure, relieve themselves, and continue their stroll almost uninterrupted. Also called a "whiz stand."

urinate   To pee.

water closet   The invention of the water closet marks man's final ascent to civilization. The concept of a water closet dates back to ancient civilizations. The Palace of Knossos on the island of Crete contained a room that used the gravity of falling water, flushing human waste into a stream below. Yet it took the genius of Sir John Harington to invent the first water closet with workable parts. Three hundred years later, British inventors perfected the water closet, or toilet as we Americans know it. The name water closet comes from the small room in which the device was kept. The French liked to refer to it as the *closet l'anglaise*. Prior to the WC, the closet may have held a chamber pot for the same use. Initially, the public disapproved of the WC as being indecent. Nowadays we would not survive long without it.

The late nineteenth and early twentieth century versions included the more modernized wash-down and the wash-out closets, which continued to rely on gravity to flush the water through.

The term water closet eventually came to refer to a *room* that contained a toilet rather than to the toilet facility itself. Names we now know the WC by include: toilet, closet, comfort station, comfort room, commode (a term popular in the southern States), potty, cloakroom, wash-down closet, john, jacks, latrine, lavatory, used beer department.

REFERENCES

Addyman, P. V. "The Archaeology of Public Health at York, England." *World Archaeology* 21, no. 2: 244.

Ariès, P., and G. Duby. *A History of Private Life: Revelations of the Medieval World.* Cambridge, Mass.: Harvard University Press, 1988.

Ashburn, P. M. *The Elements of Military Hygiene.* 1915.

Baglin, D. *Dinkum Dunnies.* Dee Why West, N.S.W.: Eclipse, 1971.

Beecher, C. *A Treatise on Domestic Economy for the Use of Young Ladies at Home and at School.* New York: Harper, 1852.

Bornoff, N. *Pink Samurai: Love, Marriage and Sex in Contempary Japan.* New York: Pocket Books, 1991.

Bourke, J. G. *On the Border With Crook.* New York: Scribner's, 1891.

Boyce, C. *Dictionary of Furniture.* New York: Roundtable, 1985.

Briggs, A. *Victorian Cities.* London: Odhams, 1963.

Camesasca, E. *History of House.* New York: Putman, 1971.

Carcopino, J. *Daily Life in Ancient Rome.* New Haven, Conn.: Yale University Press, 1940.

Catton, B. *Picture History of the Civil War.* New York: American Heritage, 1960.

Chapman, R. "A Stone Toilet Seat Found in Jerusalem in 1925." *Palestine Exploration Quarterly* 124 (January/June 1992): 4–8.

Chevalier, J., and A. Gheerbrent. *A Dictionary of Symbols.* Cambridge, Mass.: Blackwell, 1993.

Collins, G. "Potty Politics: The Gender Gap." *Working Woman* 18 (March 1993) no. 3: 93.

Committee on Science, Space, and Technology. *Contract Management Issues: Cost Overruns on NASA's Shuttle Toilet.* Government Printing Office: 1993.

Da Vinci, L. *The Notebooks of Leonardo Da Vinci,* arranged and translated by Edward MacCurdy, vol. 2. New York: Reynal and Hitchcock, 1938.

Davis, W. S. *A Day in Old Rome: A Picture of Roman Life.* New York: Biblo and Tannen, 1963.

Dawood, N. J., trans. *The Koran.* New York: Viking, 1990.

Dixon, D. M. "A Note on Some Scavengers of Ancient Egypt." *World Archaeology* 21 (1989) no. 2.

Donno, E. S. *Sir John Harington's A New Discourse of a Stale Subject, Called the Metamorphosis of Ajax.* New York: Columbia University Press, 1962.

Dubois, Abbe J. A. *Hindu Manners, Customs and Ceremonies*. Translated by Henry K. Beauchamp, C.I.E. from 3rd ed., 1906, Delhi: Oxford University Press, 1978.

Duby, G. *A History of Private Life: Revelations of the Medieval World*. Cambridge, Mass.: Harvard University Press, 1988.

Dulton, R. *Chateaux of France*. London: 1957.

Edwards, N. *The Archaeology of Early Medieval Ireland*. London: Batsford, 1990.

Elliott, C. *Princess of Versailles: The Life of Marie Adelaide of Savoy*. New York: Ticknor and Fields, 1992.

Erikson, E. H. *Young Man Luther: A Study in Psychoanalysis and History*. New York: Norton, 1958.

Fannin, J. W. *Johnnies, Biffies, Outhouses, Etc.* Burnet, Tex.: Eakin, 1980.

Florin, L. *Backyard Classic: An Adventure in Nostalgia*. Seattle: Superior Publishing Co., 1975.

Ford, J. H. *Elements of Field Hygiene and Sanitation*. Philadelphia: P. Blakiston's Son and Co., 1917.

Forman, B. M. "Furniture for Dressing in Early America, 1650–1730," *Winterthur Portfolio* 22: 149–64, Summer/Autumn, 1987.

Gamerman, A. "An Urban Archaeologist's Manhattan Privy Mystery." *Wall Street Journal*, 24 August 1993, sec. A, p. 12.

Gaunt, J. "A Year to Savour—At Least for Weirdness," *Reuter Library Report*, 27 December 1991.

Geismar, J. H. "Where Is Night Soil? Thoughts on an Urban Privy." *Historical Archaeology* 27, (1993), no. 2:57–70.

Givens, R. with K. Springen. "Splish, Splash, It's More Than a Bath," *Newsweek*, 5 May 1986; 80–81.

Gooderham, G. H. "The Passing of the Outhouse," *Alberta History* (Winter 1992): 9–11.

Guerrand, R. *Les Lieux: Histoire des commodites*. Paris: Editions La Decouverte, 1985.

Hamlin, T. *Architecture Through the Ages*. New York: G. P. Putman's Sons, 1940.

Hardy, Q. "We Can Laugh, But Once Again Japan Has Forged Ahead of Us." *Wall Street Journal*, 10 November 1992, sec. B, p. 1.

Harington, J. *The Metamorphosis of Ajax*, reprinted from original editions, edited by Peter Warlock and Jack Lindsay. London: Franfrolico, 1927.

Harris, M. *Privies Galore*. Wolfeboro Falls, N.H.: Sutton, 1990.

Hawke, D. F. *Everyday Life in Early America*. New York: Harper and Row, 1988.

Heurgon, J. *Daily Life of the Etruscans*. New York: Macmillan, 1964.

Horwitz, T. "Endangered Feces: Paleo-Scatologist Plumbs Old Privies." *Wall Street Journal*, 9 September 1991, sec. A, p. 1.

Humes, J. C. *The Wit and Wisdom of Benjamin Franklin*. New York: HarperCollins, 1995.

Jansen, M. "Water Supply and Sewage Disposal at Mohenjo-Daro," *World Archaeology* 21 (1989), no. 2.

Joyce, P. W. *A Social History of Ancient Ireland,* vol. 1. Dublin: M.H. Gill, 1920.

Keefer, F. R. *Military Hygiene and Sanitation.* London: Saunders, 1918.

Kelly, C. B. *Best Little Stories From World War II.* Charlottesville, Va.: Montpelier, 1989.

Kilroy, R. *The Compleat Loo: A Lavatorial Miscellany.* London: Victor Gollancy, 1984.

Kira, A. *The Bathroom: Criteria for Design.* Ithaca, N.Y.: Cornell University Press, 1966.

Kluger, J. "Patently Ridiculous," *Discover* (December 1992): 50–53.

Kolanad, G. *Culture Shock! India.* Portland, Ore.: Graphic Arts Center Publishing Co. Times Ed., 1994.

Lambton, L. *Temples of Convenience.* New York: St. Martin's, 1979.

Lawrence, A. W. *Greek Architecture.* Harmondsworth, U.K.: Penguin, 1957.

Longford, E., ed. *The Oxford Book of Royal Anecdotes.* Oxford University Press, 1989.

Lutske, H. *The Book of Jewish Customs.* New York: Jason Aronson, 1986.

MacCurdy, E., ed. *The Notebooks of Leonardo Da Vinci.* New York: Reynal and Hitchcock, 1938.

Maddock, T. *Pottery: A History of the Pottery Industry and Its Evolution As Applied to Sanitation With Unique Specimens and Facsimile Marks from Ancient to Modern Foreign and American Wares.* Philadelphia: Thomas Maddock's Sons, 1910.

McNeil, I. *Joseph Bramah: A Century of Invention 1749–1851.* New York: Augustus M. Kelley, 1968.

McNeill, W. H. *Plagues and Peoples.* New York: Doubleday, 1977.

McPhee, P. *A Social History of France, 1780–1880.* London: Routledge, 1992.

Miller, P. C., and R. Willock. *Continental Cans, Etc.: A Tourist's Guide to European Plumbing.* New York: Kanrom, 1960.

Mittman, K., and Z. Ihsan. *Culture Shock! Pakistan.* Portland, Ore.: Graphic Arts Center Publishing Co. Times Ed., 1991.

Morris, J. *A Winter in Nepal.* London: Rupert Hart-Davis, 1963.

Morse, E. S. "Latrines of the East," *The American Architect.* Reprinted 18 March 1893.

Muir, F. *The Frank Muir Book: An Irreverent Companion to Social History.* London: Heinemann, 1976.

Muller, K., ed. *Brussels: Insight City Guides.* Singapore: APA Publications, 1992.

Munan, H. *Culture Shock! Borneo.* Portland, Ore.: Graphic Arts Center Publishing Co. Times Ed., 1988.

Nylander, J. C. *Our Own Snug Fireside: Images of the New England Home 1760–1860.* New York: Alfred A. Knopf, 1993.

Opie, I., and M. Tatem, eds. *A Dictionary of Superstitions.* New York: Oxford University Press, 1989.

Ordronaux, J., M.D. *Hints on Health for the Use of Volunteers*. New York: Appleton, 1861.

Partridge, E. *A Dictionary of Slang and Unconventional English*, 7th ed. New York: Macmillan, 1970.

Planning, O. M. *Toilet Laughs*. Tokyo: Toto Publishing, 1992.

Quindlen, A. "A (Rest) Room of One's Own," *New York Times*, 11 November 1992, sec. A, p. 2..

Reid, D. *Paris Sewers and Sewermen*. Cambridge, Mass.: Harvard University Press, 1991.

Reisner, R. *Graffiti*. New York: Cowles, 1971.

Reyburn, W. *Flushed With Pride: The Story of Thomas Crapper*. London: MacDonald, 1969.

Reynolds, R. *Cleanliness and Godliness or The Further Metamorphosis*. Garden City, N.Y.: Doubleday, 1946.

Rice, E. *Captain Sir Richard Francis Burton: The Secret Agent Who Made the Pilgramage to Mecca, Discovered the Kama Sutra, and Brought the Arabian Nights to the West*. New York: Scribner's, 1990.

Richburg, K. "Flushed With Pride in Hong Kong." *Washington Post*, 29 May 1995.

Rivet, A. L. F., ed. *The Roman Villa in Britain*. New York: Praeger, 1969.

Roach, M. "What I Learned From Dr. Clean . . . ," in *Reader's Digest* (February 1995): 61–64.

Shepherd, C. "News of the Weird," *The State Journal-Register* (Springfield, Ill.) 24 March 1995, sec. A, p. 11.

——"News of the Weird," *Fresno Bee*. 31 December 1994, sec. A, p. 2.

——"News of the Weird," *Fresno Bee*. 30 July 1994, sec. A, p. 2.

——"News of the Weird," *Star Tribune*. 28 October 1993, sec. E, p. 11.

——"News of the Weird," *Star Tribune*. 1 April 1993, sec. E, p. 13.

——"News of the Weird," *Star Tribune*. 18 February 1993, sec. E, p. 7.

Sinclair, K., with I. Wong Po-Yee. *Culture Shock! China*. Portland, Ore.: Graphic Arts Center Publishers, 1990.

Spears, R. A. *Slang and Euphemism*. New York: Jonathan David, 1981.

Stevenson, J. *The Life and Death of King James the First of Scotland*. Glasgow: Maitland Club, 1837.

Tenenbaum, D. "Sludge." *Garbage* (October/November 1992).

Tindall, B., and M. Watson. *Did Mohawks Wear Mohawks? And Other Wonders, Plunders, and Blunders*. New York: Morrow, 1991.

Turner, E. S. *The Shocking History of Advertising*. England: Penguin, 1968.

United States Army Infantry School. *Introduction to Field Sanitation*. Fort Benning, Ga.: U.S. Government, 1984.

Vespucci, Amerigo. *Letter to Piero Soderini, Gonfaloniere*. Translated by George Tyler Northup. Princeton, N.J.: Princeton University Press, 1916.

Wallechinsky, D., and I. Wallace. *The People's Almanac*. Garden City, N.Y.: Doubleday, 1975.

Wallechinsky, D., and I. Wallace. *The People's Almanac #2.* New York: Morrow, 1978.

Ward, A. "Seattle: Bawdy Past, Perfect Present," *Forbes American Heritage* (April 1994): 70–88.

Weber, W. J. *The Unflushables: Outhouses, History and Humor.* Indianapolis, Ind.: Weber, 1989.

Wheeler, B. E. *Outhouse Humor: A Collection of Jokes, Stories, Songs, and Poems About Outhouses and Thundermugs, Corncobs and Honey-dippers, Wasps and Spiders, and Sears and Roebuck Catalogues.* Little Rock, Ark.: August House, 1988.

Wilton, T. "Potty Theories." *New Statesman and Society 7* (1 April 1994), no. 296, p. 26.

Wolfe, T. *The Right Stuff.* New York: Farrar, Straus Giroux, 1979.

Wood, T. *What They Don't Teach You About History.* New York: Derrydale, 1990.

Wright, L. *Clean and Decent: The Fascinating History of the Bathroom and the Water Closet and of Sundry Habits, Fashions and Accessories of the Toilet Principally in Great Britain, France, and America.* London: Routledge and Kegan Paul, 1960.

Yin, S. M. *Culture Shock! Burma.* Portland, Ore.: Graphic Arts Center Publishing Co. Times Ed, 1994.

Zhu, J., ed. *China: Toilet Needs to Be Revolutionized,* 1988.

——"The Toilet: A Celebration." *Harper's* (July 1993).

——"Is Elvis Dead? If So, What Killed Him? Verdict Is Due Today." *Wall Street Journal.* 29 September 1994, sec. A, p. 1.

——*The Encyclopedia of Classical Mythology.* New York: Prentice-Hall, 1965.

ACKNOWLEDGMENTS

I would like to thank all my family and friends for enduring countless conversations at the dinner table on the toilet's history. Their loyalty and stomachs went beyond the call of duty. Specifically I'd like to acknowledge those who assisted me with translations: Betsy Braden (French), Laura L. (Chinese), and Fusae and Tetsushie Okawa (Japanese). In addition, I could not have completed this project without the support and editing skills of those closest to me: Melissa Hudson, Sandy Muraca, Hal Horan, Fatima Rodrigues, Robert Grossman, Robert and Ingrid Bolding, the Perrys, Eileen Dunn, Justine Hunter, and Tim Wells. And of course, thanks Mom!

INDEX